The Physicists and God

Anthony Van den Beukel

The Physicists and God

The New Priests of Religion?

Genesis Publishing Company
North Andover, Massachusetts

215,
B.E 2

Translated by Dr John Bowden from the Dutch
De dingen hebben hun geheim. Gedachten over natuurkunde, mens en God.

© 1990 Uitgeverij Ten Have, Baarn, The Netherlands
First published in English 1991 by SCM Press Ltd, London

ISBN 1-886670-01-3 softbound

Library of Congress Catalog Card Number: 94-74212

Amer. advisor: Ernest H.Hofer, Professor emer., Amherst, MA
Cover design by Susan Hey, Gorchev & Gorchev, Woburn, MA
This book was printed and bound by Courier, Lowell, MA.

Published by *Genesis Publishing Company, Inc.*,
1547 Great Pond Road, North Andover, MA 01845-1216.
Tel. 508 688 6688; Fax 508 688 8686.

775

Contents

Introduction

For many people science is a word that conjures up feelings of awe and makes them shiver. They remember the subject from their schooldays as a world difficult to get into, the finer points of which they never understood. By taking pains and working hard they could get fifty or perhaps just scrape sixty in their tests, but this was more a reward for their zealous toil than an assessment of the depth of their insight. They also remember that there were always one or two boys (and sometimes a girl) in the class who with no apparent difficulty could keep getting ninety or even one hundred in the subject. They understood it. A ten for science (or mathematics) was something very different from a high mark for French, let alone music or drawing, subjects which didn't count at all. *The brightest students in the class* - that's the halo that scientists wear around their heads. Who wouldn't love to be the father or mother of such a clever child? Of course, it's also good to have a nice child, but niceness doesn't get you very far. Brightness does.

Science and scientists are held in awe because they are generally thought to be clever. This recognition is perhaps not based primarily on the fact that scientists are capable of constructing complicated theories which are almost impossible to explain to lay people, who express amazement at the bits and pieces that they can grasp or think that they can grasp. There is something else which gives these theories a solid, convincing basis which is evident to everyone: the *results* that they produce, the marvels of technology. Radio, television, computers, nuclear energy, compact discs, spacecraft, and so on. That is really awesome. If there are people who can master formulas for achieving all this, what can't they do? Can't they, for example, also provide the basis for a world-view? Can't they show us our place in the universe, a way to the deeper meaning of existence?

Such questions would be forcefully answered in the negative by the great majority of scientists, but not by all of them.

There are also those who enjoy being gurus and show as much in popular books which are sold by the million and read avidly. One of them, whom we shall meet again later at length, puts it like this:

> The tremendous power of scientific reasoning is demonstrated in the many marvels of modern technology. It seems reasonable, then, to have some confidence in the scientist's world-view also.

I mentioned another emotion which many ordinary people experience when confronted with science: I said that it makes them shiver. It seems to be a cool, not to say chill, world of apparatus, figures, and formulas. Little warmth emanates from it. It's an "objective" world. The universe is a collection of microscopically small particles which developed at the very beginning from a Big Bang. After that, matter and radiation spread over immeasurable space. Somewhere in a corner, on an unpretentious satellite of an inconspicuous star, human life developed. Scientists can describe very precisely how all this came about. In a book called *The First Three Minutes*, one of them has described the events which must have taken place in the whirling, intensely hot soup during the first three minutes after the Big Bang. It makes breathtaking reading, a feast for the understanding, but one's heart remains cold. What are human beings really? Merely a collection of atoms dominated by the laws of nature?

For believers, there is a further problem here. Where is God in all this? Didn't God create the world? Can't traces of God be found anywhere? "No," say some contemporary scientists, "you can forget God. God doesn't exist." "As a physicist you have to have a split personality still to be able to believe in a God," said a recent winner of a Nobel Prize for science. His remark was made in a newspaper interview. "We're wasting our time here," kids of fifteen tell their pastors or Sunday School teachers. "God doesn't exist. That's been proved by science." Evidently, they too read the paper. Many other scientists - I think, the majority - don't go so far as that (but because they don't make pithy comments they don't get into the newspapers). "We don't know," they say. "We don't encounter God in our investigations, but we can't prove that God doesn't exist either!" Yet others accept that this is the case at present, but expect that things will be different within the foreseeable future. Science is approaching its end: Before long we shall understand "everything." If God is part of that "everything," he will make an appearance himself. If there is a God, we shall catch him. "That," says one scientist, "would be the ultimate triumph of human reason, for then we would know the mind of God."

In this book, we shall investigate the basis of such expectations. We shall begin by first taking a look around the everyday world of scientists. What sort of people are they? Cool, objective seekers of truth, above the earthly bustle? Gurus? Or perhaps people to whom nothing human is alien? What are they occupied with? At least it is clear that they are trying to understand the physical reality around them and to sum it up in clear, compact, elegant formulas. But what kind of a reality are they occupied with? Is it "objective," i.e. independent of the observer who perceives it? Is it the *whole* of reality? If scientists ever finish their work - and some people think that this will happen before long - will they then understand *everything*? (And what does "understand" mean?) Or does science describe just an aspect of total reality, and are there, as Hamlet says to Horatio, "more things in heaven and earth than are dreamt of in your philosophy"? And is God part of that "more things in heaven and earth"? Where is God, and how is God to be found? Can anything be proved about God? And what do we mean by "prove"?

This book will go into that kind of question, but not in a predominantly abstract, theoretical way. We shall be listening to *people*: great scientists, like Newton, Pascal, Einstein, Hawking. I have also brought my own experiences into it. I have more than thirty years of teaching science and scientific research behind me. During that time I have tried to be a believing Christian. I shall consider whether the two can go together, while preserving one's integrity. This is - also - a *personal* account. That can be a weakness: It may also prove its strength. You, the reader, must judge for yourself.

1 It's Cold in Delft

"I have sold myself body and soul to science, being in flight from the 'I' and the 'we' to the 'it'"
Albert Einstein

Kees Andriesse is a physicist who wrote poems and composed string quartets in his youth. He is an expert on nuclear energy and is in favor of building new nuclear power stations. At the same time, he is a convinced Social Democrat and a member of the Democrat Party.

This dry summary contains obvious tensions. It doesn't take much imagination to see him busy convincing a sympathetic audience of his views about the peaceful uses of nuclear energy. Poems and string quartets do not easily share the atmosphere of lecture rooms where science is taught and Reason is uppermost. That can't work out well.

Nor did it work out well; or at least things did not go smoothly and without tensions. In 1985 Kees was on the edge of a nervous breakdown and, on the advice of his doctor, went for a week to the island of Terschelling. He was blown about in November, when storms and squalls shriek over the island, which by then had been abandoned by all the tourists. There he thought, or rather was overwhelmed by thoughts, feelings, and questions, about his existence as a scientist and a human being and the relation between the two. The book in which he described this experience he called *A Boudoir on Terschelling*. Here is a passage from it.

When I went to Delft in 1958, it was as though I was being thrown into a tank of cold water. All the ornamentation was stripped off; it was measuring and counting. This is 1 and that is 1 and 1 + 1 = 2. The rest didn't matter. The rest was stripped off. The rest was uninteresting junk, rubbish. Just as you have to let crystals grow gently, by slow cooling, from the molten mass or the vapor, so that all the atoms can find an orderly place in them, so you must let the ingredients of human beings settle down gently so that they can live in harmony and without inner tensions. But if they cool down quickly, structural faults, brittleness and instability occur. So, at the beginning of my study I was squashed into an unstable glasslike mass instead of a stable, crystalline Kees.

When I read that, my student days, which I, too spent at this faculty in Delft, were already thirty-five years behind me, but I recognized the description as though I had experienced it yesterday. In the meantime, I have discovered that Andriesse and I are not the only ones to look back on our scientific education with such feelings. Since I have begun to refer to these things now and then in my lectures, I have been able to note from the response of the students how many of them still feel the same chill in their bodies even today.

Where does it come from? Are lecturers in science, or scientists generally, a special kind of person, void of human emotions, for whom the whole of existence is reduced to measuring and counting, to mathematical formulas, to a closed and rational whole? That is certainly not the case. Anyone who gets to know scientists and becomes one of them soon discovers that just like other human beings they can be sad and enraptured, can be fond of music, love their wives, and lie awake over their children. That's not the point. The point is that scientific lecturers who enter a lecture room put aside all this for a couple of hours and present themselves to their audience as those for whom the world is organized clearly and rationally, who fill the board with incontrovertible equations. Often unintentionally, this gives the impression that everything outside this (Andriesse's "the rest") is not worth mentioning. The suggestion is that one can tackle a problem, *any problem*, by analyzing it in a cool, critical, objective way and arrive at a solution by logical reasoning. That must happen in the case of scientific problems, but there are risks when it becomes an attitude of life. It can lead to a personality structure in which the balance between feeling and understanding is disturbed at the expense of feeling. Andriesse calls that "structural faults," with a metaphor derived from solid-state physics.

The danger is far from imaginary. I assumed above that the emotional life of natural scientists is no different from that of other people. However, that is questionable. I know a minister near Delft whose duties include the pastoral care of members of his congregation. In modern times, an important part of this work involves dealing with marriages which have gone wrong. The pastor has had so much experience in this that he can divide them into different categories. One of them he calls "the Delft marriage." In this marriage "he" studied at the Technical University in Delft, and "she" is an ordinary, normal woman. He is proficient in logical argument and in any difference of opinion uses this skill to "prove" that he is right. He is always right, because she cannot get a word in edgeways, not so much because the argument is incontrovertible, but because she lacks the training to point to the weak point in it. Nevertheless, on the basis of her feeling, she refuses to yield. "There's more to it than that."

Things may go like this. She is worrying about a problem which is very important to her. She has to choose between two alternatives, but can't make up her mind. He listens to her and says, "Oh, is that all? That's no problem. Let me prove it to you. There are two possibilities, A and B. Each has something to be said for it. If there is a lot to be said for A and not much for B, the decision is easy. It gets more difficult, the less there is to be said for A and the more there is to be said for B. The most difficult decision is when there is as much to be said for A as for B. But precisely in this case it doesn't really matter much which you choose. So, the most difficult decision is at the same time the easiest. There is no such thing as a difficult decision." And he picks up his newspaper, because another of the world's problems has been solved.

You shouldn't think that this is an extreme example. When my wife heard the story about the Delft marriage, she said, "I can recognize a lot in that." Years ago, I myself thought up the demonstration about difficult decisions, and since then I've applied it many times, to my complete satisfaction, when I've had to take a decision. The overvaluing of the understanding at the expense of the heart is a professional deformity which few practitioners of the exact sciences wholly escape. It was expressed more than three hundred years ago by the great mathematician and scientist Blaise Pascal (1623-1662) as follows:

> I spent a long time in the study of the abstract sciences, and was disheartened by the small number of fellow-students in them. When I commenced the study of man, I saw that these abstract sciences are not suited to man, and that I was wandering farther from my own state in examining them, than others in not knowing them.

I shall come back later at length to what Pascal means by "my own state," i.e being human. Here I shall just venture this short and incomplete description: To be human is to stand open with heart and understanding to the world, to fellow human beings and to God. We shall not encounter human beings and God in the natural sciences. They are carefully and deliberately excluded. As we shall see, a philosophy lies behind this.

In this connection let's look at the way in which scientific publications are organized and the language in which they are written. Any article in a technical scientific journal is broadly divided up as follows. The first section is called "Introduction," and it sets out the purpose of the investigation. The problem is formulated, and the way in which it is to be tackled is indicated very directly. In the second section, "Experiments," there is a description of the apparatus used and an indication of the method of measurement. The third section is called "Results" and gives the results obtained, often in the form of diagrams and tables. In section 4, "Discussion," the results are compared with existing theoretical models, and if they do not fit in with these models completely, an attempt is made to adjust the model or sometimes to suggest a new one. In the last section, "Conclusions," which is as brief as possible, what has been achieved is summed up very concisely. In short: 1. This is what we want. 2. We do it like this. 3. It comes out like this. 4. It is to be understood like this.

In most cases, this description of the research is only remotely connected with what actually took place. Often the researcher began elsewhere, with another aim and with other expectations than the result that was actually achieved. Some approaches proved dead ends, there were disappointments, and suddenly new ideas emerged, new ways were sought and sometimes found. The apparatus built at the beginning did not work, or was inadequate and constantly had to be adapted and extended. The research cost blood, sweat, and tears. Nothing of that is to be found in the final result, the published article. It has been carefully purged of any odor of sweat, of any human emotion.

That is also expressed in the language used in the articles. It is English, but of a special kind. It betrays no trace of an author's personal style. You only have to read half a page of Saul Bellow or John Fowles to know whom you're dealing with; there can be no mistaking that. But the language of scientific publications shows no such identity. The person of the author remains completely outside it. The vocabulary is also very limited - no more than a few hundred words. It's robot English, language which a computer can cope with. The impersonal element is stressed further by the consistent use of the passive. Words like "I," "my," or, in the case of several authors, "we," "ours," are completely missing. There is no "I noticed" or "we concluded," but "it was observed" and "it is concluded."

For anyone familiar with this from his/her youth, it all seems the most natural thing in the world. You don't know any better. So it came as a surprise to me to discover that things haven't always been like this. In the *Transactions of the Royal Society* for the year 1672, there is an article which is thought to be among the most famous in the history of science. It is about the splitting of white sunlight into a spectrum of colors with the aid of a prism. It was written by Isaac Newton. The first lines are as follows:

To perform my late promise to you, I shall without further ceremony acquaint you, that in the beginning of the year 1666 (at which time I applied myself to the grinding of optic glasses of other figure than spherical), I produced a triangular glass prism, to try therewith the celebrated phenomena of colours. And for that purpose having darkened my chamber, and made a small hole in my window shuts, to let in a convenient quantity of the sun's light, I placed my prism at his entrance, that it might be thereby refracted to the opposite wall. It was at first a very pleasing diversion to view the vivid and intense colours produced thereby; but after a while applying myself to consider them more circumspectly, I was surprised to see them in an oblong form; which according to the received laws of refraction, I expected would have been circular.

That won't do! Here we have the words of someone, a man, who makes and keeps promises, grinds lenses, constructs prisms, and is proud of doing so (*my* prism!), who derives pleasure from attractive colors, who had expected something but was then surprised because he saw something different and writes that down openly. Nowadays the article would be rejected by any third-rate journal because of its use of unscientific language.

The following point is also revealing. Professor Van der Hoeven (of whom otherwise I have nothing but good to say) quotes the same passage in his book *Newton* in what he calls an abbreviated form. This abbreviated form consists in a literal rendering of the text given above, but with the omission of the passages with a personal color which I brought out earlier. With one exception: "I was surprised" is "translated": "It aroused my interest." Evidently that is still permissible: The researcher's interest may still be aroused, but he may not be surprised - or at any rate, he may not give any indication of that fact. So Newton's offending text is purged of subjective blemishes and made suitable for digestion by twentieth-century readers who think that they know what may be expected of science: concern with "objective" reality, detached from human elements, which may only be reported in objective terms.

It might be retorted that all this sounds somewhat exaggerated. What need is there for personal digressions in scientific literature? The journals would be significantly bigger than they already are, and far too much paper would be wasted. So it's a matter of economy. Moreover, as scientists we are always aware that the description of our research in scientific articles is an abstraction and not an account of what really took place. That last point, the involvement of personal emotions, the tortuous ways, the sweat poured out, we know well enough from our own experience. Surely that doesn't need to be repeated *ad nauseam*?

The objections are not unreasonable, though they are not very convincing. Economy with paper is not a prominent feature of the scientific world. There is paper enough for anything of any importance and much which is not important at all. Omitting the subjective passages in Newton's article only saves a few lines. Moreover, "I observed" is shorter than "It was observed." As for the second difficulty: That is precisely what happens in all novels, poems, music, and paintings. They are always about the same human (emotional) world of experience, but they don't get boring.

However, I don't want to start a case for the reintroduction of personal effusions into scientific literature. What I am concerned to do is to point out that the form and language used are *symptoms* of an attitude of mind, a philosophy. The form and content are completely one. And the content, the message: Remember that this is about an *objective reality* which exists outside human beings, but which can be known by them and set out in simple laws which can also be applied to human beings themselves.

The question whether there is such an objective reality has been a major issue in twentieth-century science. The most important, most famous, and most persistent champion in answering this question in the affirmative was Albert Einstein. In the meantime, the question has been answered, and apparently definitively. The answer is, "No." We shall look at that more closely in a later chapter.

2 Nothing Human is Alien

"We are so presumptuous that we would wish to be known by all the world, even by people who shall come after, when we shall be no more; and we are so vain that the esteem of five or six neighbors delights and contents us" *Pascal*

George Steiner, the English linguistic philosopher - once at Cambridge, England, now in Geneva - somewhere relates the following experience from his childhood years. When he was about six years old, he overheard a conversation between his father, who was a banker, and someone else.

Thereupon I asked my father what the difference was between bonds and shares or between the bank and the stock exchange. He said, "I know that so that you never need to know it." He had decided to make it possible for me to become a scholar or a writer. It was his aim in life that I should never know such nonsense and that I should never waste my time over something so idiotic.

Evidently, as Steiner senior imagined things, there were two worlds. On the one hand was the ordinary human world concerned with money, power, and influence, a struggle which usually does not bring out the most attractive human characteristics. On the other hand was the world of scholarship, in which such emotions play no role, in which, in tranquillity, dedicated scholars exclusively occupy themselves with the search for truth: an exalted world, populated by serene, ethereal academics.

The academic world of the 1950s which I entered as the shy assistant to a professor showed many parallels with the image that Steiner senior had. Its most marked characteristic was tranquillity. Not the tranquillity of the cemetery but the tranquillity, the stillness, which made it possible to reflect with deep concentration on the scientific problem with which you were occupied. This atmosphere was admirably described by Karel Van het Reve, at the time professor of Slavic languages at Leyden University, who, on being asked by an interviewer, "What does a professor do all day," replied, "Well, I sit and doze at my desk a bit, and now and then I write something." It was an academic atmosphere in which the problems being worked on were chosen by the researchers themselves in complete freedom and without any restrictions. Researchers were never asked, for instance, about the social relevance of their activities. Far less were they asked whether their "production" was large enough. When the research led to results which completely satisfied researchers, and they thought it worth while having them published in international scientific literature, then they did that. On the whole, so much self-criticism was practiced in the process that the number of publications per man-year of research was very small, but they were of good, solid quality. These publications secured the scientist access to the international forum of scientists, with whom he or she could maintain contact through correspondence and by regular visits to conferences. The exchanges there were predominantly friendly, sympathetic, appreciative.

That is how I remember it, though my recollection may be colored by the distance of years, which always has the effect of wiping out the less attractive features. On the other hand, I find the same atmosphere in the splendid memoirs of the grand old man of Dutch physics, H.B.G. Casimir (entitled *Haphazard Reality*), in which he wrote about his experiences in the 1920s and 1930s, among the great scientists of that time.

I don't want to claim that nothing of this academic atmosphere remains nowadays, but that the picture does need to be considerably qualified. What, for example, happened to George Steiner? He describes a visit that he paid in 1956, as a young student in Budapest, to the great philosopher George Lukacs.

When I entered his study, confused and awkward because I so admired him, he was sitting at his desk, a small man with a large head. Behind him stood forty volumes of his own work. He looked at me - that's how you make people uneasy - and said nothing. I had to say something. The only thing that occurred to me was, "How were you able to write all that?" His splendid answer was, "Nothing to it, Steiner. House arrest! House arrest!" Of course he was right. Now we rush from one stupid congress to another. We get 366 invitations a year. We're airport professors. At present, airports are our university. But when you're under house arrest you write, you work, you reflect.

The most important change that has taken place is not so much that our students no longer come into our rooms awkward or confused because they so admire us, but say "Hi" or at best "Hello." Steiner notes that the tranquillity has gone and that this is not conducive to concentrated thinking.

How has that come about? There is a complex of causes, and I won't mention all of them. One of the most important is the marked growth in the size of universities. Student numbers are many times what they were forty years ago. The numbers of professors and scientific workers have kept pace with them. As a result, the potential for scientific work has also increased enormously. Moreover, in subjects like physics, where experimentation has a central place, the costs of the increasingly more sophisticated apparatus are many times what they were. The total costs get so enormous that Congress and Administration have felt compelled to set limits on them. Nor is that all. They've also begun to ask questions. What's the point of all that research? What contribution is it making to society? Are all these researchers working hard enough? What are they really producing? How can you measure it?

Thus money has made its entrance into the serene world of science as an important factor, and, as Steiner senior knew from experience, that is never conducive to preserving a good atmosphere. In addition to all their other roles - teachers, researchers, members of numerous committees - present-day professors have also become entrepreneurs. They run shops. The staff, apart from the professor, usually consists of two permanent colleagues, five PhD students, a technician, and a secretary. The annual salaries of this team amount to half a million dollars and more. In addition, there is the cost of laboratory space and apparatus and central supplies, together with workshops and administration. The annual budget that I mentioned is a low estimate of the total cost of this little shop.

What is the product, which is called "output"? Predominantly academic articles which are published in specialist journals. For a group of the size indicated, these can number between ten and twenty a year. The average extent is around five pages per article, so at a low estimate the cost price is around $6000 per page. The product also has to be sold. The first thing that has to happen is for the article to be accepted by a journal of as high academic standing as possible. I shall come back shortly to what problems may be involved. But that's not all. Steps have also to be taken to see that the article is read by other scientists and if possible even quoted by them in their publications to indicate that one's work has at least not remained unnoticed.

That doesn't happen automatically, as you might think. It would be naive to suppose that the only factor which counts is the quality of published work, that if it is good, it will automatically be noticed by professional colleagues who zealously study the literature to keep up-to-date with the state of science. It is not unusual for that to happen, but it's no longer a matter of course. The growth of scientific research world-wide is explosive, and the same goes for the amount of published work. Even within the specialisms, which are constantly getting narrower, it is no longer possible to read, let alone study, everything that appears. One has to make a choice. And this choice is certainly not just limited by the quality of the article (to discover that, you have to read it first!). So, whether the name of the author means anything to you becomes an important factor. Knowing the name is the thing, and here you will immediately recognize an important concept from the advertising world. Just as a clothing manufacturer hopes that a purchaser in a shop will recognize a brand from the logo on the shirt of a famous player in a tennis tournament, so a present-day scientist hopes that a colleague who has to decide whether or not to study his article will remember his name and face from the conference at which he gave that attractive paper, asked a particularly clever question, or had a friendly chat with him at a meal. It is also good for getting your name known if you can reach that point on the international circuit, where the prestigious publicity functions, like editor of a journal or member of the organizing committee of a conference, are distributed. It is not unimportant here to have good friends in the right places.

What I have just said is loaded and one-sided. One really doesn't go to conferences just to advertise oneself, but also, and above all, to discuss common issues with colleagues and listen to their questions. You don't win a Nobel prize by enlisting the aid of a good publicity agency. The best scientists do rise to the top. But one or two levels lower, where good and valuable work is still being done, the degree to which you know how to make your way is of an importance which is not to be underestimated.

Why is it really so important to scientists for their articles to be published, read, and quoted? In the first place, of course, because that implies a recognition of their achievements by others (I shall come back to this). But in the second place, usually, down at ground level, because that is vitally necessary for them if they are to get the financing they need for the continuation of their work. This does not drop from the skies. The large annual sums needed for the average team has to be provided by an organization with that money, a sponsor. The main sponsor of scientific research at universities is the government, and after that industry. This money has to be distributed among the research teams, all of whom have significantly more requests than can be financed by the means available. That leads to pressure around the food troughs.

I needn't go into detail about the procedures by which resources are distributed. The consequences can easily be imagined without that. The results of research must constantly be assessed. For this, researchers must regularly (say, once every two years) produce an extensive report on what they have done and what the results have been. In these results, the list of publications is the most important item. In the presentation of their account, it is worthwhile for scientists to remember the rules of the publicity world. In particular, it is worth stressing the beneficial consequences that the research may have in possible practical applications. Here, too, modesty is not a virtue which brings rewards. The report is presented to a number of colleagues - usually internal - who have to assess it and give it a mark. The scale runs from 1 to 10: 1 is the highest, Nobel prize, level, and is never given. The figures actually awarded almost always lie between 3 (still very good) and 5 (respectable, but not tremendous). It can be very important whether one scores 3.4 rather than 3.9. The outcome of the marking is awaited by researchers with the same tension as students await grades.

So, colleagues are the judges. The owners of rival shops. There are risks in giving a high mark to a colleague: At any rate, the money that goes to him does not come to you. Colleagues, certainly in small countries, are people that one knows. These are people whom, as well as being more or less competent, one thinks of as being pleasant or unpleasant, arrogant or modest, threatening or innocuous. It is clear that in such circumstances it is not easy to arrive at an objective and honest judgment of their work. I am convinced that by far the majority of us try to be serious, and I am equally convinced that we do not succeed, or do not succeed enough. We, too, are people who under the pressure of the struggle for life and the survival of the fittest are not always governed by the highest ethical norms.

Now we need to look more closely at the way in which the results of research are accepted or rejected by the scientific world. *The Theft of Prometheus* by Kees Andriesse (whom we came across earlier) has an account of an experience which is instructive. The author, who was appointed as a physicist to an astronomical laboratory to be a specialist on physical measuring apparatus, had sufficient time there to do research on his own account. This was in the sphere of astrophysics, in which he could combine his knowledge of science with the astronomy that he picked up from his environment. He worked completely alone, in great freedom, but was treated with indifference by the astronomers around him, on a theoretical investigation of the physical events taking place around a remote star, Carina. Before his fascinated eyes, the result slowly took the shape of a simple, elegant formula which, as he supposed and later demonstrated, covered a wide area of observations in astrophysics. He was delighted.

And then it happened. He presented the results in the form of an article to a first-rate scientific journal. It was rejected. A second, equally famous, journal did not reject it immediately, but made difficulties over a long period. (To those who don't know this, the journals enlist referees to judge the articles which are offered them; arbitrators, knowledgeable professionals. In first-rate journals, these are heavyweights. In that case the rejection of an article amounts to being rejected by the establishment in your professional sphere.) Having got tired of this, Andriesse decided to try a stage lower. A journal of considerably less prestige than the first finally decided on publication. In the meantime, he had also presented his findings verbally to a number of scientists. However, in almost all cases he came up against a lack of interest, skepticism, unbelief. It was not until years later that he finally got a breakthrough, resulting in an "invited lecture." We shall come back to that later.

This experience of Andriesse's is not an isolated one. I myself have the strong impression - which I can demonstrate with many examples - that the difficulty of getting results accepted in the scientific world becomes greater, the more innovative, original, and pioneering they are (in retrospect!). I shall limit myself to two examples which relate to two of the greatest figures in the history of science.

Newton's work on the splitting of light with a prism, which I mentioned earlier, was in fact published immediately in the *Transactions of the Royal Society*, but that was only the beginning of his difficulties. The work was not accepted by his colleagues. Some went so far as to doubt the accuracy of the experiments, and Newton was extremely hurt. Others, like Boyle and Hooke, two prominent scientists of the time who were appointed by the Royal Society as referees to study the article critically and to report on it to the Society, accepted the experiments but rejected the explanation. In particular, Hooke thought that Newton's results did not demonstrate anything new but could very well be explained by his own - Hooke's - theory of light. (Here we come upon the paradoxical feature that the most prominent scientists may not be the most objective judges of new ideas. Didn't they achieve such eminence because their own ideas have been generally accepted? That does not make it easy for them - and here we have a common human characteristic - to accept results which stamp their own work as outdated and thus do some damage to their reputation.) The dispute between Newton and his opponents as to who was right was carried on with all due vigor and unnecessary bitterness. The controversy bitterly disappointed Newton, and his eagerness to publish his work, which was already not very great, was further diminished. (Look at the dates: 1666 discovery of chromatic dispersion, 1672 publication.) Only in 1704, another thirty-two years later, did he publish all his insights into the nature of light in his great work *Opticks*. This date is connected by some people, probably rightly, with the fact that Hooke died in 1703. Newton's antipathy to controversies was allergic.

The same thing happened to Einstein. "For his contributions to theoretical physics and especially for his discovery of the law of the photo-electric effect" was the reason given by the Nobel Prize Committee which awarded him the prize - finally! - in 1921. The discovery mentioned was made in 1905. His special theory of relativity, by which he became much better known to the general public, and which in itself deserved a Nobel prize, dates from the same year. But even in 1921 the committee had not gotten that far. Still more astonishing is what is said in the reasons given by the committee, and above all what is not said there in connection with the photo-electric effect. This is the phenomenon that electrons can be released from a piece of metal by directing light on it (it is used, among other things, for camera light meters). The gist of the law formulated by Einstein, which is mentioned by the committee in the reasons which it gives for the award, is that the energy of the electrons released is proportional to the frequency of the light falling on the metal. This was accepted by everyone at the time on the basis of experimental results. Not so, however, the quite revolutionary *explanation* given by Einstein for this law. Even in 1921 that was still doubted unanimously by all the leading physicists.

Einstein's explanation was that a beam of light, at least in this experiment, must be imagined as a stream of *particles*, light quanta, later called *photons*. Revolutionary, because not so long before, there had been a controversy lasting two centuries over whether light is a wave phenomenon or a stream of particles, and this controversy seemed to have been decided definitively in favor of waves. Here Einstein was taking a step backwards again. Even worse, he did not deny that light was (also) a wave phenomenon. His explanation inferred that it was not one *or* the other, but one *and* the other. That was beyond all imagining and was therefore unacceptable. Only some years later (from about 1925) did it have to be accepted by everyone under the overwhelming weight of the evidence. It formed the basis for the really revolutionary discovery of twentiet a-century science, quantum mechanics. Abraham Pais, the author of the brilliant biography of Einstein (*Subtle is the Lord*), gives a gripping description of the history of the photon, from rejection to acceptance. I shall not follow it here, but I shall take one more illustration from it.

In 1913 Einstein was nominated as a member of the Prussian Academy of Sciences by four prominent physicists, including Max Planck. In their letter of commendation, they praised him at length as a brilliant physicist, but ended with the following excuse for his mistakes: "That he may sometimes have missed the target in his speculations as, for example, in his hypothesis of the light quanta, cannot really be held too much against him." In connection with this story Pais observes: "Physicists are conservative revolutionaries, resisting innovation as long as possible and at all intellectual cost, but embracing it when the evidence is incontrovertible." We shall see below how - ironically enough - Einstein himself in later life became a splendid illustration of this statement because he was utterly opposed to the acceptance of quantum mechanics, of which he was one of the principal founders.

You may perhaps ask yourself what these stories are meant to show. "All right, the acceptance of new ideas takes time and trouble, resistance has to be overcome, but in the end the truth triumphs. The physicists you've discussed were still able to celebrate their triumph royally in their lifetimes. Things are different elsewhere in life. Wasn't the German engineer Diesel dead long before anyone saw anything in his engine? Didn't Vincent Van Gogh die a pauper without ever having sold a single canvas?" That's not an unfair comment in itself, though here I have described, exclusively, successful stories with a happy ending. We have absolutely no idea how many brilliant scientific ideas never get anywhere, and because journals have no archive of rejected articles, this history will remain unwritten. But that is not the subject of this chapter. Here I am concerned to sketch out the world of science as a world of people to whom nothing human is alien. I want to bring out one more aspect of this.

Say to a farmer, "What splendid work you're doing, producing all this milk for the good of humanity," and he will look at you as if you're crazy. He's never thought about that for a second - as far as he's concerned it's self-evident. The meaning of his work is a fact - his only concern is whether he can live on it. Say to a scientist, "I read your last article, I thought it particularly good," and he or she will be grateful to you. Scientists need recognition of the meaning of their work. The need for recognition is greater, the more doubt there is about this meaning.

There are a number of reasons for this doubt. One criterion for the success of an article is the regularity with which it is quoted by others in the scientific literature. Fifty per cent (half!) will never be quoted at all. There is no tangible indication that they have been read. The work will be available for inspection for a number of years in libraries in the form of thick volumes of periodicals bound in leather, a volume per year. In the course of time they will be transferred to the stacks. The other half are on average quoted less than twice (1.7 times, to be precise). That, too, is not very good and cannot form any firm basis for a feeling of fulfillment on the part of the average researcher. The quotations usually die out after five to ten years. Researchers can at least hope, and now and then suppose, that these articles have contributed something to the progress of science. They can measure this to some degree by the extent to which the work has a place in standard works on the subject. Only a very small proportion stand the test of time, then at the level of the Nobel prizewinners.

In this light we can see the value of the following passage. Andriesse, whose contributions to astrophysics finally achieved recognition when he was invited to give a lecture, describes what happened.

And then there was my lecture, after so many others. Applause, long-drawn-out applause, but no questions. I just walked round afterwards, and liked that. Then a little woman from the society approached me. She said, "I know what men are like; they don't say what they think. My husband works with Conti in Boulder. He's already had your article at home for quite a while. When he'd read it, he was silent. "Here's a guy who's understood," he said. Then he made a long distance call from home to a Harvard colleague who had also read your article. The conversation lasted almost an hour. Do you know how expensive a long-distance call is?"

He doesn't say whether he embraced her. I don't think that there are many researchers who do not recognize very clearly the feeling conveyed in that quotation.

One last thing. I've associated the question of the meaning of research with the resonance that it finds among others. That's important; it's an elementary human need, but it's not the only thing. For those involved in scientific research, that research also has a meaning in itself: It creates moments of great beauty, deep satisfaction, a feeling of happiness. These moments are not frequent, but they can make up for long periods of tension and setback. They can also keep researchers going who have not found any response. At least as long as the sponsors keep on supporting them!

3 Understanding Everything

"A complete, consistent, unified theory is only the first step: our goal is a complete *understanding* of the events around us and of our own existence"
Stephen Hawking

IN SEARCH OF THE SUPERFORCE

Simplicity is a characteristic of the truth. That is a proverb which was not invented by scientists (it comes from classical antiquity) but is written on their hearts. Scientists are seekers of simplicity, and that search is rewarded. "The world," Newton wrote, "which to the naked eye exhibits the greatest variety of objects, appears to be very simple in its internal constitution, and so much simpler by how much the better it is understood." The simplicity of the nature which researchers encounter is written in the language of mathematics. Simple, elegant mathematical formulas accommodate a world of phenomena. For anyone who has eyes for it, these formulas radiate beauty. Hence scientists are often fond of repeating Keats' line, "Beauty is truth, truth beauty."

For the scientist, the creation of order out of a confusion of phenomena, which at first sight have nothing to do with one another, is simplicity. The tidal flows of the oceans, the fact that an apple falls from a tree, the movement of the planets round the sun, seem to be three completely independent phenomena. It was Newton who demonstrated that all three are based on the same law, the universal law of gravity, which has an extraordinarily simple and elegant form. The law states that there is a force of attraction between two bodies which is based on "mass" and which diminishes as the distance between masses grows greater. The force of gravity works between the moon and the waters of the ocean, between the earth and an apple, between the sun and a planet. The force of gravity is extraordinarily weak compared with the other forces which occur in nature (about which I shall be saying more shortly), so that it is only clearly perceptible when at least one of the bodies has very great mass. In the examples I mentioned that is the case: The moon, the sun, the oceans, and the planets are very "massive" bodies. Therefore, the force of gravity is dominant in the world of the stars, but cannot be detected in the world of atoms. Other forces are dominant there.

So, there are electrical and magnetic forces. Electrical forces are forces between two electrically charged particles, of which there are two kinds, those with a positive and those with a negative charge. The names of these charges are not arbitrary, because when they are brought together the two kinds of charge cancel each other and can together produce a zero charge. Thus the atom of the simplest element, hydrogen, consists of a nucleus with a positive charge, a proton, around which a much lighter particle with a negative charge, an electron, circles like a planet round the sun. The charge of the electron is as strongly negative as that of the proton is positive. The total charge of the atom is thus zero, and the atom is electrically neutral. Two particles with positive charges exert a power of repulsion on each other: So too do two particles with negative charges, whereas a particle with a negative charge and one with a positive charge attract each other. The law which describes the force between two charged particles is just like Newton's law of gravity. But the "mass" which occurs in the latter is now replaced by the charges of the two particles.

It is difficult to explain precisely what magnetic forces are. They are forces between two magnets, but that does not shed much light on matters. Two things are clear. First, that a magnet always has two "poles," a north pole and a south pole, which have opposite kinds of "magnetic charge." Two north poles (of different magnets) and two south poles repel each other, and a south and a north pole attract each other. Second, a magnet can only be made of a very limited number of kinds of materials, of which iron is the best known. But we are not concerned here with a more detailed understanding of electricity and magnetism. What concerns us is that in the course of the nineteenth century it became increasingly clear that the two are connected in some way. A wire through which an electric current runs seems to exert a force on a magnet in its environment. Conversely, a magnet moving in a coil of wire can produce an electric current (that is the basis for a bicycle dynamo, for instance). After a long history, Maxwell ultimately succeeded in bringing the two phenomena together completely under a single denominator: the theory of electromagnetism. The mathematical formulation of this theory, Maxwell's four equations, is a miracle of beauty, of compactness, and of concentrated power of expression. The equations describe the whole extended sphere of electromagnetic phenomena such as x-rays, light, and radio waves (which were forecast by Maxwell's equations and only discovered later).

Maxwell's theory is a splendid example of *unification*: the combination of two or more groups of phenomena which apparently have completely different characters, bringing them under a single denominator. Apart from the two basic forces of nature which we have met so far, gravity and electromagnetic force, there are two others. And it can easily be demonstrated that there must be at least one more. Earlier we encountered the hydrogen atom, consisting of two parts: a proton with a positive charge as a nucleus, and an electron with a negative charge circling round it. The force which keeps these two together is electromagnetic force. All other atoms are more complicated. Thus a carbon atom has a nucleus consisting of six protons and six neutrons (a neutron is a particle with almost the same mass as the proton, but without an electrical charge). Another six electrons circle round these, which remain bonded to the nucleus of the atom by the electromagnetic interaction. But what about the nucleus itself? This contains six protons with a positive charge which repel one another. If that were all (at this level the force of gravity does not play a part), the nucleus would burst apart. There must therefore be a force which holds the nucleus of the atom together. And that is indeed the case. It is called strong nuclear force. Alongside that, within the nucleus of the atom another force plays a role which is necessary to explain, among other things, the phenomenon of radioactivity. That is the weak nuclear force.

Such was the position until recently. There are four basic natural forces: the force of gravity, electromagnetic force, and strong and weak nuclear force, each of which is at work within its own sphere and is well, or increasingly better, understood. Now all scientists hope that this picture is too complicated. They dream of the possibility of a further unification, just as happened earlier with electrical and magnetic forces. The most attractive solution would be if all four forces were to prove manifestations of one single natural force, superforce.

That is no longer a dream. Since the 1960s, great progress has been made towards a unification of the four natural forces. Nevertheless, the road is far from being at an end, but some eminent physicists believe that complete unification will be achieved within ten years. The name of the book in which the English scientist Paul Davies tells the fascinating story of developments over the last twenty years is *Superforce*. It reads like a thriller. It is impossible for us to follow all the story here, but we should pause over one particular part of it: the first step, the successful attempt to reconcile two of the four forces, namely electromagnetic force and weak nuclear force, to make what is now called electroweak force. The two theoretical physicists who succeeded in doing this, quite independently, in 1967, were Steven Weinberg and Abdus Salam.

The Weinberg-Salam theory is too complicated to set out briefly here, but one aspect of it can be explained; it is also important for a later stage of our story. An old question with which Newton was already concerned when he arrived at his law of gravity was this: How are the natural forces really transmitted from one particle to another? For example, the law of gravity says that the sun and the earth exercise a force on each other, but space between the two is largely empty. How can this force then be felt over such an immense distance? Is it telepathy? The answer to this question has become clear only in the course of our century. It is not telepathy; the working of the force comes about by an exchange of particles, so-called "messenger particles." Each of the four fundamental forces has its own kind of messenger particles. Thus the messenger particles of electromagnetic force are the photons which we came across earlier. Those of gravity are called gravitons, but for the moment they are hypothetical particles, because they have never been observed, and the chance that this will happen in the foreseeable future is thought to be very small.

It follows from the Weinberg-Salam theory that three messenger particles, called W+, W- and Z particles, must be involved in weak nuclear force, and the theory predicts what these particles must look like. Although the confidence of physicists in the theory was so great that Weinberg and Salam won the Nobel prize for it in 1979 (along with Glashow, who had done the preparatory work), the demonstration of the three messenger particles in an experiment would be a very welcome, indeed indispensable confirmation of the theory.

Such an experiment was carried out successfully at the Institute for High Energy Physics of the European Community, CERN (Centre Européen des Recherches Nucléaires) at Geneva in 1983. The gigantic size of this Institute can be measured by its annual budget: around $600 million, the cost of a thousand university teams. Hundreds of scientists and technicians of very high standing work there. The apparatus is awesome. There are machines which can accelerate particles like protons and electrons to fantastic values, close to the speed of light. The greatest accelerator available at CERN in 1983 was the proton-antiproton accelerator, extending over many miles (in the meantime a new and far bigger machine has already been completed). This was the machine which made possible the experiment for demonstrating the W and Z particles. For very high energies are needed to "liberate" these particles so that they can be observed. In this case it was expected that the experiment would succeed if a beam of protons could be made to collide at high speed with a beam of antiprotons (particles with the same mass as the proton, but with a negative charge).

The man who played a decisive role in the success of the experiment was Simon Van der Meer, a Dutch physicist, who in 1953 graduated from the Delft Faculty of Applied Physics and had worked at CERN since 1958. He broke through the bottleneck of the experiment, the production of a beam of antiprotons of sufficient intensity, by developing a marvelous piece of apparatus, a combination of a cunning physical principle and engineering skill of a high order. A year after the world had come to hear of the success of the experiment, Simon Van der Meer, along with the leader of the research group, Carlo Rubbia, was awarded the Nobel prize for science. This was a thoroughly deserved, undisputed reward for a technical and scientific achievement, the glory of which also reflected a little on the Netherlands and the Delft faculty which produced him.

INTERVIEW WITH A NOBEL PRIZEWINNER

When a scientist who has been used to doing his work in tranquillity, far removed from the world of publicity, wins the Nobel prize, his tranquillity comes to an end. He cannot stop the flood of invitations to give lectures, has to accept honorary doctorates, and the media, television, newspapers, and journals crowd round his door asking for interviews. That is what happened to Simon Van der Meer.

The curiosity of the journalists goes further than the scientific achievements by which the person concerned has won fame. What sort of a person is he? What are his views on human life? Has that anything to do with his science? I know of three interviews with Van der Meer. From them, he seems to be a modest person. "A machine builder," he calls himself, and dismisses any comparison with great scientific thinkers of the calibre of Einstein. He is fond of gardening, carpentry, and the music of Bach. He relates that he is descended from a Protestant family of simple faith.

To begin with he felt at home and secure in that faith, but already in his youth, he rejected this security as a form of self-deception, and since then he has regarded all forms of religious belief as expressions of a fanaticism from which much - if not all - evil has flowed. His tone, otherwise so calm, becomes fierce when he talks about this.

I have taken the following remarks from an interview in the *NRC-Handelsblad*, the Dutch equivalent of *The New York Times* (April 18, 1987; the italics are mine):

> The meaning of his work? It is purely concerned with knowing. The principle of life can be *understood.*
> One day it will even be possible to build a machine like a human being.
> What we are finally trying to discover is: Why is *everything* as it is?
> As a physicist, you must have a split personality if you're still going to be able to believe in a God. To look for religious explanations is to put the problem in the wrong place.
> You look for a theory which explains *everything* in a logical, elegant way. To think that this exists is *perhaps also a kind of faith.* It is the most basic thing you need for *understanding the world.*

A photograph is printed alongside the interview. In it, down the full length of the page, you can see the scientist standing on a pedestal. In his right hand, which is raised slightly above his head, he is holding up a globe. The photograph indicates that this is not causing him the slightest difficulty: In his hand the world is small and light. The photograph is in complete harmony with the content of the interview. It is a work of art.

The picture immediately had two associations for me. "He's got the whole world in his hand" is a line from a Negro spiritual. "He" is God, who in the photograph is replaced by science (the scientist). "We have made science our God," says Casimir somewhere, and that is what the photograph expresses. In that faith there is no room for faith in other gods, certainly not in the "Old One" of former days. A physicist who nevertheless makes the attempt has a split personality.

A second association the photograph had for me was the picture which depicts the figure of Atlas from Greek mythology. Here, too, we see man and the world, but in a totally different relationship. It shows tiny man, bent double under the burden of an almost unbearable, leaden world. His name, Atlas, expresses his position: the one who suffers intensely. We have indeed made progress since then. The photograph of the scientist shows us the human being in triumph, the master, the ruler. Here is autonomous man, who has things under control, who no longer has the world on his back but has literally taken it in hand and lifts it up like a feather. He has a straight back and does not bow or kneel to anyone. He is the man of Progress.

The instrument by which progress has been made is science. The task of science is to understand, *to understand everything*. Everything consists of matter and the forces which the particles of matter exert on one another. To know this is "the most basic thing you need" for understanding the world. If you can fathom that, there are no more mysteries. That this "all" does *not* mean reality *in so far as* it is accessible to methods of physical measurement and physical theory is clear from the context. It is even clearer from the words of Stephen Hawking, the English physicist whom we shall meet later, which I have put at the head of this chapter. Hawking, whose ideas show considerable agreement with those of Van der Meer, has no difficulty in associating superforce with "the events around us" and "human existence." Everything will be understood. Emotion. Sorrow, anxiety, joy. The sonnets of Shakespeare. Bach's cantatas.

It's an old story, the origin of which can be identified precisely. In a book about world history, in the chapter about the eighteenth-century French philosophers of the Enlightenment, I read the following passage:

The father of French materialism was Frederick II's physician, Julien Offrat de la Mettrie. His studies of medicine and biology led him to explain the expressions of the human spirit as purely physiological facts. He declared that thought is a function of the brain and human beings differ from animals in that their brains are more refined.

The title of his main work, *L'homme machine* (Man as a Machine), is typical of his standpoint. Since in his view what we call spiritual life is one of the functions of the body, the soul can no longer exist when the body dies. Belief in an immortal soul is thus nonsense. Everything can be explained from the standpoint of materialism; the hypothesis of the existence of a God is therefore superfluous because it can lead to fanaticism and dispute. Despite their hatred of Christianity, most materialists were really idealists, more fighters for a belief than thinkers. They believed in Progress; they had an unshakable trust in the capacity of the Enlightenment to overcome tyranny and oppression, superstition and fanaticism, evil and sin.

The quotation is an almost literal rendering of Van der Meer's views.

Let's now pause for a moment over the question what "understand" really means in modern science. By way of contrast, I shall first give an illustration of what it meant in classical science. It is well known that a quantity of air enclosed in a container exerts pressure on the walls of the container. This pressure can be measured with a manometer. If the volume of the container is reduced by half, for example with a piston, the pressure of the gas doubles. This observation was developed into a formula in the seventeenth century by the English scientist Robert Boyle as follows: For an enclosed quantity of gas the volume times the pressure is constant. That is a law, a description of a group of observations for which an understanding is sought. That can happen as follows. Suppose that the gas consists of a collection of particles, atoms, which move at great speed through the enclosed space and in so doing also regularly collide with the wall. In this collision, they make a tap on the wall which represents a force. That is the cause of the pressure. If one reduces by half the space within which they move, they collide with the wall twice as often and the pressure becomes twice as great. This can also be calculated mathematically with the help of Newton's laws of motion, and that produces Boyle's law. The calculation gives another interesting result: The constant value of pressure times volume is connected with the average velocity of the atoms. That can also be understood directly: The more quickly the atoms move, the more powerful will be the tap that they make and the greater the pressure of the gas. And that is again connected with another observation: If one keeps the volume of the container the same but increases the temperature of the gas by heating it a bit, the pressure of the gas increases. Now we also understand that temperature is directly connected with the average velocity of the atoms. The connection can be expressed very precisely in a formula.

The description given here has two characteristics. The first is that anyone can "understand" the *model* used, billiard balls flying around and striking one another and the wall, because it corresponds directly to our capacity for imagination. Abstract concepts like pressure and temperature are clarified in a way which anyone can understand. We feel very satisfied with this. The second characteristic is that the description of what happens in the container is given in terms of mathematical equations.

In modern science only the second characteristic is left. At the level of atoms one has to forget one's capacity for imagining, which is grafted on to our experiences in the large-scale world around us. Many people (and clandestinely still many scientists) imagine the hydrogen atom as a particle consisting of a small, spherical nucleus, around which another much smaller particle, the electron, circles like a planet round the sun. That is now a prohibited, or at least a meaningless, conception of things. We have a mathematical equation, Schrödinger's equation, in which entities appear like the "wave function" and the "energy levels" of the electron and which, if it is solved properly, makes predictions about what can be observed on that electron, for example the wavelength of light that a hydrogen atom can transmit when it is "excited." These predictions then fit the observations admirably, and modern physicists are completely satisfied with them.

So, in modern science, to "understand" means exclusively to describe observed physical phenomena with the help of an abstract mathematical formula. In a strictly materialistic sense, to "understand everything" means that everything, including "the events around us" and "human existence," can be described in mathematical terms. If we get this far, then the paradise, the kingdom of heaven of materialistic belief, has dawned. Whether such a "blessed" state has been achieved, even Stephen Hawking seems to doubt when he reaches the last page of *A Brief History of Time* (my italics):

> Even if there is only one possible unified theory, it is just a set of rules and equations. What is it that breathes fire into the equations and makes a universe for them to describe? The usual approach of science of constructing a mathematical model cannot answer the questions of why there should be a universe for the model to describe. Why does the universe go to all the bother of existing? Is the unified theory so compelling that it brings about its own existence? *Or does it need a creator, and if so, does he have any other effect on the universe?* And who created him?

One last remark. A world in which everything is understood seems to me to be an extremely discouraging prospect. We can see why from what has been happening to chess in recent years. Chess is now played not only by human beings but also by computers. The power of computers to play chess is steadily increasing. Computers can already beat Masters and sometimes Grand Masters. The computer programmers are full of confidence that within the foreseeable future the computer will become a world champion, perhaps even that the result will be the one, definitive chess match: White plays and wins. That would be a triumph of calculation, delegated by human beings to computers which can be better than they are. If that should happen - which is still the question - then chess will be fully understood. At the same time it will be stone dead, murdered. There will be no need to play chess again. Could the symbolism of the photograph of the Nobel prizewinner with the globe not also be this: The world has already become so small because we understand so much of it? If we understand everything in it, it shrivels away to nothing. There is no need to experience it.

I think that the chances of our seeing this are not great, either with chess or with the world. The Grand Masters of chess do not seem to be losing much sleep over it. They seem to think that chess at their level has an incomprehensible element that exceeds the computer's power of reasoning. Call it inspiration, intuition, spirit. As far as the world is concerned, perhaps in the not too distant future the day will dawn when the theory of superforce is complete, when a simple, elegant mathematical description will be found for all observed physical phenomena - in principle. That will be a great day for the scientists, and above all for those who are directly involved in carrying through the masterwork: a triumph of human ability, of the power of human thought. But it will not really change the human world.

In the autumn of 1969, I spent a period in the United States. It was three months after the Americans had put a man on the moon for the first time. I met a young scientist there who was deeply disillusioned. He had, he said, expected that after this fantastic achievement of human reason, defying all imagination, humankind would change, that an age of brotherhood would dawn. Nothing had happened. I listened to him open-mouthed in amazement. You have to be an American to be like that, I thought, so childlike, so naive. I felt European, old and cynical. I doubt now whether this sort of excessive expectation about the contribution of science to the welfare of humankind is so typically American. But I think that those who cherish it will go away with the same flea in their ear.

When Maxwell took the first step towards the unified theory, Hitler and Stalin had yet to be born. When Weinberg and Salam had begun on the second step rather later, there were Pol Pot in Cambodia and the Gang of Four in China with their mass killings. The third and fourth steps will pass just as unnoticed in this respect. Humankind probably needs something different if it is ever to become anything. "The most basic thing that you need" is more than a formula about superforce. Perhaps it is indeed a creator who, in Hawking's words, "has yet other consequences for the universe." And, I would add, for human beings in particular.

4 The End of Objective Reality

"We are not merely spectators. We are actors in this great drama of nature" *Niels Bohr*

THE QUANTUM REVOLUTION

When some years ago Stephen Hawking took up his position as professor at the University of Cambridge in England (where he had the same chair as Isaac Newton three centuries before), as usual he gave an inaugural lecture. The title was "Is the End in Sight for Theoretical Physics?" His answer, based on developments in the theory of superforce, was cautiously affirmative: It is quite possible that physics is almost "finished." It is worth noting that precisely a century ago, towards the end of the nineteenth century, the same comment could be heard quite widely. Pieter Zeeman, later a Nobel prizewinner, was fond of telling how in 1883, when he had to choose what to study, people had strongly dissuaded him from studying physics. "That subject's finished," he was told, "there's no more to discover." It is even more ironic that this also happened to Max Planck, since it was he who, in 1900 precisely, laid the first foundations for one of the greatest scientific revolutions in history, the quantum revolution.

What was "finished" towards the end of the nineteenth century, a more or less well rounded whole, is what we now call classical physics. For the most part, this covered three areas: mechanics, the theory of motions and forces, the basis of which had been laid by Newton; electromagnetism (including optics, the theory of light), which had been given its definitive formulation in Maxwell's laws; and heat theory (thermodynamics), with which names like Clausius and Kelvin were associated. These theories were moving into what Casimir calls the "technical stage": The foundations were established, so that little that was new could be expected in this area (this was what Zeeman's advisers meant). Work in physics consisted of developing constantly new applications on the basis of these foundations. However, we should not belittle this. Even in our days, a large number of physicists are still active unearthing enough new discoveries from the gold mine of classical physics to give them complete satisfaction in their work. But there is nothing new in the deepest, fundamental sense, and that is the only thing that "frontline theorists" of Hawking's calibre are interested in.

The characteristic of the classical laws of nature which is important for the discussion in this chapter is that they are *causal and deterministic*. What is meant by this can be clarified by an illustration. Let's recall the container filled with atoms of gas, from the previous chapter. At a particular moment each atom, envisaged as a small billiard ball, will be in a particular position. Each will also have a velocity, the magnitude and direction of which we should be able to indicate with a little arrow. We call this situation the initial state. If the velocities of two atoms in the cluster are such that these meet each other later, there is a collision. At the time of this collision, they exercise forces on each other, which change their velocity. If we can know these forces, we can calculate by Newton's laws the velocity of these two atoms after the collision. In principle (!) we can do that for all the atoms in the group. Starting from the initial state, we can calculate the complete history of every atom into the distant future. That is completely fixed; it is *determined*. By the strict laws of nature, the causes (the forces) determine the consequences (the changes in velocity) of every atom. There is a *causal* link between the two.

What holds for the group of atoms in the container can easily be extended to all the atoms of which the universe is composed. So, there is little cause for surprise when one comes upon the following famous statement by the French mathematician and scientist Pierre Laplace, from the beginning of the nineteenth century:

An intelligent being who at any moment knows all the forces in nature and also all the positions of all the things of which the universe consists, should be in a position to reproduce the movements of the greatest bodies *and those of the smallest atoms* in a single formula; supposing that this being were in a position to analyze all events, *then nothing would be uncertain for him*, and both future and past would be open to him.

The sting lies in the words in italics. We shall see that uncertainty is fundamentally located precisely in the world of the smallest atoms. That apart, it is not at all surprising that Laplace, when asked by Napoleon where the place of God was in his picture of the world, replied, "Sire, I have no need of that hypothesis."

That was the situation at the end of the nineteenth century: Almost all known natural phenomena could be described satisfactorily with the aid of a limited number of basic laws of nature. I say almost, for even then there were already a number of things which did not fit, but these were regarded as clouds in an otherwise clear sky. In the first decades of the twentieth century, the clouds were rapidly to increase in number and extent to form a thundercloud. We have already met one of them (in Chapter 2): Einstein's view that what was observed in the photo-electric effect compelled the conclusion that light must have the character of particles as well as of waves. A second serious problem emerged when in the first decade of the twentieth century the structure of the atom began to become much clearer.

In classical physics, the atom still played a minor role. In some theories, like the kinetic theory of gases which I have described, the atom was introduced as a hypothesis: Atoms were regarded as hard, solid little spheres which could collide with one another like billiard balls. At the end of the nineteenth century, the "atomic hypothesis" was still disputed; even then, many physicists still did not believe in the existence of atoms. That was soon to change, especially as a result of the famous experiments of Ernest Rutherford (1907), which for the first time created a picture of the internal structure of the atom. The experiments showed that an atom had to be regarded as a kind of planetary system: Around an atomic nucleus with a positive charge, a tiny sphere, electrons with a negative charge (much tinier spheres) circled like planets round the sun. The atom consisted for the most part of empty space. Now this picture was in direct conflict with Maxwell's classic laws of electromagnetism, which predict that such an atom cannot be stable, but must collapse.

We shall not follow here the exciting events which took place between 1900 and 1925. Nor shall we investigate how at this time Newton's classic laws of mechanics and theory of gravity were replaced by Einstein's revolutionary theories of relativity: the special theory of relativity (1905) and the general theory of relativity (1915). Here I shall note only that around 1925 the thundercloud of phenomena in the atomic sphere which could not be fitted into the classical laws was replaced by a completely new conception, that of quantum physics. We shall now look at its most prominent features.

The first characteristic of the natural laws of quantum physics is that elemental events at the level of atoms have a *statistical* character. You cannot predict precisely when and how a particular event will take place; all you can say is something about the probability, the *chance*, that something will happen. That can be demonstrated by an illustration, the radioactivity of atomic nuclei. Some atomic nuclei, like those of a particular kind of thallium, are unstable, and can disintegrate. In this process the thallium nucleus emits an electron of high velocity, and what remains is the nucleus of the element lead. The electrons released can be captured, for example on a photographic plate, so that their number can be counted. From such experiments it can be established that of a piece of thallium which contains billions upon billions of atoms, after about four minutes half the atoms will have disintegrated. In the following four minutes, another half of the remaining atoms will disintegrate, so that there is then only a quarter left, and so on. This "half-life time" can be measured precisely. This "macroscopic" behavior, the behavior of a large collection of atoms, can be predicted very precisely with the formulas of quantum mechanics. But things are different when we look at an "elementary event," in this case the disintegration of one atom. Nothing can be predicted about that: It can happen within a second, after a day, or over a thousand years.

A game of dice is a precise comparison here. The outcome of one elementary event, a single throw of the dice, cannot be predicted. But we know that the chance of throwing a three is a sixth, by which we mean that if we throw six thousand dice, the number of times that we throw three will be close to a thousand. The greater the number of throws, the more accurate the prediction will be. Evidently, the forces which are at work within a radioactive atomic nucleus also work like this and cause the outcome. It must be stressed here that physicists believe that this does not happen because we know too little about these forces, so that for the moment we have to resort to the "dice description" until we have become clever enough to predict the behavior of each individual atom precisely. Rather, it lies in the fundamental nature of the forces that they are like dice, and as a result *it is in principle impossible ever to know more than we do now.*

Of course this means that *causality* at the level of atomic events has gone. The same cause can have infinitely many different effects. Of two atoms which are no different from each other, one can disintegrate after a thousandth of a second and the other after ten thousand years.

The second characteristic of quantum physics is the wave-particle duality. We already came across that in an earlier chapter on light. In the nineteenth century it seemed to have been established definitively that light is a wave phenomenon. On the basis of observations of the photo-electrical effect, Einstein concluded that light also has the character of a particle, a view which he maintained in isolation for twenty years, but which around 1925 had become inescapable. However, Einstein himself did not like this. "There are therefore now two theories of light," he wrote, "both indispensable and - as one must admit today despite twenty years of tremendous effort on the part of theoretical physicists - without any logical connection." Things were to get even worse. In 1924 the French scientist Louis de Broglie hypothesized that the converse could also be the case, that the "things" which we had hitherto regarded as particles, like electrons, could possibly have the character of waves. He himself predicted what the "wavelength" of such a particle would have to be. It was not long before this idea, too, was confirmed by experiments. Now it is the most natural thing in the world. We investigate details of matter invisible to the eye with the aid of light waves in a light microscope or with electronic waves in an electron microscope.

So, around 1926 the twofold, dual character of matter and radiation were beyond doubt. This is a duality which was (and is) beyond our understanding because it is evidently inadequate at the level of atoms. The bold step which was taken by the founders of quantum mechanics, the most important of whom are Erwin Schrödinger, Werner Heisenberg, Niels Bohr, Max Born, Wolfgang Pauli, and Paul Dirac, was as follows. We do not wait until we can imagine one or the other, but we develop a mathematical form of description which incorporates the wave-particle duality and which is capable of predicting the results of experiments for us. One of the first and most famous results was Schrödinger's equation. Applied to an electron, the two most important entities in this equation are the energy levels and the "wave-function" of the electron. For the electron, for example, these energy levels in a hydrogen atom can be calculated with Schrödinger's equation and then used to predict what kind of light such an atom can emit in an "excited state." The result seemed to be in amazing agreement with what had long been observed in the past. Each energy level had a particular "wave-function." What did that signify?

Niels Bohr and Max Born played a major part in answering this question. Their conclusion was that the wave function says something about the probability, the chance of encountering an electron somewhere. In principle it can be anywhere, but the chance of finding it at a particular place when you make a measurement is great when the wave function is great. Just as, to use a rather feeble illustration, you can meet people in bathing suits anywhere, but the chance of doing so is greater on the beach than on the street.

Since then the success of quantum mechanics has been brilliant. Applied on a wide scale to atomic phenomena, it always produces the right answer. There is no doubt that it works. What more could you want? "Nothing" is the answer given by Bohr and by far the majority of physicists with him. The two aspects, wave and particle, are, to use a term of Bohr's, complementary; they supplement each other. Sometimes an experiment shows us the particle aspect and sometimes the wave aspect. That we cannot combine these two in one coherent conception is a consequence of the limitations of our capacity for imagination, but we must live with that. The most important thing, and this is our concern, is that we can describe what we observe and that has proved extremely successful.

The third and perhaps the most far-reaching and shattering result of quantum physics was Heisenberg's *uncertainty relations*. Let's remember Laplace's "intelligent being." If this being were to know at a particular moment the position and velocity (or, to put it more accurately, the momentum that is the product of mass and velocity) of all particles in the universe, and also the forces at work between the particles, then past and future would no longer hold any secrets for him. Now the first of Heisenberg's two uncertainty relations states that this is impossible even for one particle. If you want, you can ascertain the *position* of a particle with increasing accuracy, but if at the same time you want to measure the *momentum*, the result is increasingly inaccurate. If you know the precise position of the particle, then you don't know about the momentum, and vice versa. The reason is that in order to make an observation, observers must intervene in the particle they are observing. The slightest possible intervention that the observer has to make is to project one quantum of light on the particle. In this way it is possible to determine the location of the particle, and the lower the wavelength of the quantum of light, the more accurately this can be done. However, when the quantum of light encounters the particle, it disturbs the momentum of the particle in an unpredictable way, and the lower the wavelength, the more strongly it does so.

It is clear that all this has far-reaching consequences. Determinism is ruled out. So is the idea of "objective reality." At best one can continue to believe that there is an objective reality outside us characterized by particles which occupy a well defined place and have a well defined momentum, but for the moment that must remain a belief. If quantum mechanics has the last word, it is in principle inaccessible to experimental confirmation. There is no verifiable reality outside us independent of our perception. The percipient and the perceived are one and indivisible. (That was also the case in classical physics, with the difference that an accurate correction could be introduced for the intervention of the observer. In quantum mechanics that is no longer possible.)

That is, if quantum mechanics has the last word. How certain is that? Could it not be possible that the theory merely represents an intermediate stage and that the fundamental limitations which it seems to impose on our capacity to know will one day be removed? We shall return to these questions in more detail later.

THE EMOTIONS OF THE MAIN FIGURES INVOLVED

The main aim of this book is not to give a complete account of twentieth-century physics which is understandable to outsiders. That has already been well done by many others. I am more concerned with the people who are involved in the subject, what inspires them, and how they regard the results of their physics. So here we shall look at a few remarks by the pioneers of quantum physics, from which it may emerge how they experienced the transition to the new physics.

But before we do that there is something else. The heading above the previous section, "The quantum revolution," would prompt Casimir's disapproval, as is evident from the following passage from his book *Haphazard Reality*:

> In the introduction of new theories, many concepts must be brought in, so radically different from the images of the older theories that these themselves appear in a new light. That is attractive and exciting, but it is not a real revolution, at least not as long as we regard physical theories as an approximate description of a limited number of physical phenomena which in their turn are only a limited part of our human experiences.
>
> However, this gradual evolution of theories will be regarded as a revolution by those who have attributed unlimited validity to a theory and then make it the basis of a complete philosophy of nature, indeed even a world-view. Scientific revolutions are not made by scientists. They are only conjured up afterwards, and often not by the scientists themselves but by philosophers and historians of science.

So, this is not a revolution, but gentle evolution. It is partly a question of words. What is a "radical transformation of images" if not a revolution? And incidentally there is also this: The triple relativizing of scientific knowledge as "an *approximate description* of a *limited number* of physical phenomena which in their turn are only a *limited part* of our human experiences" speaks to me from the heart. Physicists like those we heard in the previous chapter could have learned a great deal from this. Nor can enough warnings be given against making a scientific theory the basis of a world-view. Laplace's determinism also, perhaps above all, found a great following outside science. That does not alter the fact that on one point I disagree with Casimir: The rise of quantum mechanics was not made into a revolution by philosophers and historians of science after the event, but was felt to be a revolution by the most prominent people involved, while it was actually happening.

The reactions of a number of great scientists to the totally new view which inescapably pressed itself upon them from their science, and their unwillingness and reluctance to accept it, in particular their reluctance to bid farewell to the causal-deterministic character of the laws of nature, always makes me think of the description given in the Gospels of the events of the first Easter morning. That Sunday morning, the third day after Jesus had died on the cross, signs began gradually to appear to his grieving followers that something mysterious was going on. Some reported that when they got to the rock tomb they found it empty and that they met "angels" there who said, "Why are you looking for the living among the dead? He is not here, he is risen." The striking thing about the stories is that as these people relate their experience to their friends they come up against a wall of unbelief. The others utterly reject them and seek explanations in terms of the old classical theory: "When you're dead, you're dead. If the tomb is empty, the body must have been carried off somewhere." They are convinced only if they are confronted with irrefutable evidence. The last who has to believe is Thomas, who maintains to the end that his friends could tell him even more. The stories undermine the view which has been widespread since Freud, that belief is a matter of human beings projecting their own ideas and images outwards, so that God and heaven are illusory, that human beings believe only what they want to believe.

It was roughly like this with the scientists who were confronted with the consequences of the new theory, with Einstein in the role of Thomas. Here are a few quotations (Pais' biography of Einstein is again a rich source):

Schrödinger (to Bohr): "I might not have published my papers, had I been able to foresee what consequences they would unleash."

Heisenberg: "I remember discussions with Bohr which went through many hours till very late at night and ended almost in despair; and when at the end of the discussion I went alone for a walk in the neighboring park, I repeated to myself again and again the question: Can nature possibly be so absurd as it seemed to us in these atomic experiments?"

Heisenberg again: "I myself have thought so much about these questions and only came to believe in the uncertainty relations after *many pangs of conscience.*"

Einstein: "Quantum mechanics is very impressive. But an inner voice tells me that this is not yet the real thing. The theory produces a good deal, but hardly brings us close to the secret of the Old One. I am at all events convinced that He does not play dice."

That's no small matter. Despair, absurdity, pangs of conscience. Doubt as to whether the publication of articles which won you the Nobel prize was right. And all for the same reason: that the causal-deterministic character of the laws of nature had to be given up. That was, as Pauli put it, the price that had to be paid for the new insight. Einstein was the most stubborn in unbelief. For the first five years (until around 1931), he was indefatigably occupied in devising "theoretical experiments" which would necessarily undermine the credibility, the consistency of the new theory. As in everything else, he was also brilliant here. In particular he made things hot for Bohr. Pais gives a gripping account of an almost successful attempt by Einstein, at the sixth Solvay Conference in 1930, to demonstrate that he was right. The theoretical experiment which he produced was so ingenious that Bohr could not repudiate it on the spot. Only the next morning, after a sleepless night, did he succeed. The refutation was done with the help of Einstein's theory of relativity.

After that Einstein gave up the struggle, in that he recognized that quantum mechanics is a consistent theory which does not contain any intrinsic contradiction. From then on his opposition took another form. Quantum mechanics might not be incorrect, but it was incomplete; it could not be the last word. To demonstrate this, in 1935 he conceived a "theoretical experiment" which was so ingenious and so difficult to carry out that it was fifty years before it could be translated into a real experiment. We shall be coming back to it shortly.

One more comment to end this section. I have never been able to understand properly the almost heart-rending pain and difficulty that it cost these great physicists to bid farewell to the causal-deterministic character of the laws of nature. What attraction can there be in a picture of the universe which consists of a collection of atoms which perhaps once, in the beginning, were created and set in motion by God and which since then have moved in a way that is completely prescribed by iron laws of nature, with no possibility of change? That becomes even more oppressive if one believes that human beings, too, including the human spirit, are subject to this. Then any human freedom is impossible. The consequence then can only be a fatalistic attitude to life, comparable to what was produced in theology by Calvin's doctrine of predestination, a doctrine which has never spread much sunshine in human existence. Why should you want to believe something like this, certainly if it is no longer necessary? The only explanation I can think of is that the majority of physicists believe that the causal-deterministic laws do not also relate to the human spirit, but exclusively to that *part* of reality, physical reality, with which they are concerned. In that case the uncertainty relations mean only that a limit is set on what physicists can come to know, and that is never of course a good thing. But I find this explanation unsatisfactory because it does not make the vigor of the reactions sufficiently clear.

THE DECISIVE EXPERIMENT?

That, then, was the situation at the beginning of the 1930s. Quantum mechanics worked and was generally accepted, even by Einstein. With this difference, that by far the majority of physicists, at Bohr's instigation, were convinced that the uncertainty, the "vagueness" of the theory was definitive, that it could never be got round, while Einstein thought that this situation must be temporary and that one day in the future the next step would be taken, which would again open up the way to objective reality.

"That's all very well," you might say. "That's always the case, and the future will show who is right." The fascinating thing about what now follows is that there is no need to wait for the future because an experiment is possible which will determine who is right. This experiment, which at the time still had to remain a "theoretical experiment," was proposed in 1935 by Einstein along with his colleagues Boris Podolsky and Nathan Rosen. The explanation of their idea which now follows may be too complicated for some readers; in that case they should turn over the page and read the conclusion.

We have seen that the uncertainty relations make it impossible to determine the position and the momentum of a particle with great precision at the same time. The cunning plan of Einstein and his colleagues was, nevertheless, to do that, and to do so by bringing in the help of a second particle. Let us call these two particles 1 and 2 respectively. Together they form what we call a "closed system," in other words there is no influence on them from outside. In that case it is true, for quantum particles also, that the sum of the momenta of both particles does not change in the course of time: It is constant. We now determine the momentum of the two particles separately; that can be done with the greatest possible precision, even according to quantum mechanics. Some time later we determine the momentum of particle 1. Because the total momentum has remained constant, the momentum of particle 2 can be precisely calculated at that moment. At the same moment we also determine the position of particle 2. That can also be done with all desired precision. We then know both the position and the momentum of particle 2 and the uncertainty relation is thus obviated. One more thing must also be taken care of. The intervention of the measuring of the momentum of particle 1 may not have any influence on particle 2. Given that such "influence" can come about only through a signal which according to Einstein's theory of relativity can at most be transmitted with the speed of light, the measurement must be so short that in this time a signal emitted by particle 1 cannot reach particle 2. That is no small requirement. If the particles are, for example, at a distance of one meter from each other, this means that the measurement cannot last longer than a billionth of a second. It is perhaps superfluous to stress that here too there is still no problem for the approach of quantum mechanics. The particles are not "localized." As long as we do not know its place, particle 1 is as if it were everywhere, including the place that we have established with the help of the measurement for particle 2. The measuring of the momentum from particle 1 can then exercise a direct influence on particle 2, without the intervention of a signal.

For Einstein himself, this argument seems to have been so compelling that as far as he was concerned there was no need for an experiment to be carried out. He called the possibility of quantum-mechanical influence, as described above, "spooky" or "telepathic": It was too absurd to be taken seriously. It was not until 1982 that the definitive experiment, cast in a particularly sublime form, was carried out by the French physicist Alain Aspect and his colleagues. The result was clear: The "telepathic" interaction between the two particles so detested by Einstein was demonstrated unambiguously. *It meant the end of objective reality*. (For completeness, I should also mention that for a correct interpretation of the Aspect experiment another important theoretical barrier had to be overcome, and this was done by the CERN physicist John Bell, whose name must be mentioned in this connection.)

Is the argument now finally settled? Would Einstein, if he were still alive, finally give in, or would he think up a new, brilliant refutation? Or would he just, as earlier, say, "It simply cannot be true; it conflicts with my scientific intuition?" We shall never know, but we can get some inkling from the following story which Paul Davies relates in his book *Superforce*:

Several months after Aspect published the results of his experiment, I had the privilege of making a BBC radio documentary programme about the conceptual paradoxes of quantum physics. The contributors included Aspect himself, John Bell, David Bohm, John Wheeler, John Taylor, and Sir Rudolph Peierls. I asked all of them what they made of Aspect's results and whether they thought that commonsense reality was now dead. The variety of answers was astonishing.

One or two of the contributors felt no surprise. Their faith in the official view of the quantum theory as enunciated long ago by Bohr was so strong that they felt the Aspect experiment merely provided confirmation (albeit welcome confirmation) of what was never seriously in doubt. On the other hand, some were not prepared to leave it at that. Their belief in commonsense reality - the objective reality sought by Einstein - remained unshaken. What would have to go, they argued, was the assumption that signals could not travel faster than light. There must be some "ghostly action at a distance" after all.

You rub your eyes in amazement. A group of outstanding physicists, including a Nobel prizewinner, turns out to be a group of *believers*. Those who hold the one belief do not need any confirmation; they knew it all along. Those who hold the disputed belief are unshocked by what seems to be an overwhelming argument for the prosecution and are ready to throw overboard one of the most fundamental principles of the whole of science, as if it were nothing. We shall meet one of the latest, David Bohm, at greater length in the next chapter. He too adopts a solitary position in science, but his view is well worth considering.

If outstanding, great scientists already look at the results of their science in this way, how much more seriously must we take Casimir's warning against the "attribution of unlimited validity to scientific theories, making them the basis of a complete philosophy of nature, indeed of a view of life." Paul Davies, for example, falls into this trap, and he is not the only one. He rightly notes that the human spirit was completely banished from the deterministic view of the laws of nature *à la* Laplace.

In attempting to reduce all systems to the activities of simple components, some scientists came to believe that mind is nothing but the activity of the brain, which is nothing but a pattern of electrochemical impulses, which in turn is nothing but the motion of electrons and ions. This extreme materialistic philosophy reduces the world of human thoughts, feelings, and sensations to a façade.

And then he concludes:

The new physics, by contrast, restores mind to a central position in nature [...] Common sense may have collapsed in the face of the new physics, but the universe that is being uncovered by these advances has found once more a place for man in the great scheme of things.

That may be so. The human mind, human thoughts and feelings, are back again. That is conceded because the scientists have discovered uncertainty relations. If you're not careful, God may soon be back too!

MODEST AND HUMBLE. AN AFTER-DINNER ADDRESS

During conferences at which scientists meet one another to report their progress and exchange thoughts about their ideas, one evening is always reserved for the conference dinner. The high point of the evening is usually the after-dinner address, a speech given by one of the great men of science, a "grand old man" who gives a personal view, preferably full of witticisms and anecdotes on developments in the subject. Anyone invited to give such a speech really has gotten somewhere.

At a conference on the physics of disordered systems

which was held in Israel in 1987, this honor fell to the Israeli physicist Max Jammer. The speech is printed at the end of the proceedings of the conference in *Philosophical Magazine*. Because its content follows on so perfectly from the discussion in this chapter, I have included part of the speech. In particular the ending seems to me to be a worthy conclusion to what we have been concerned with here.

Jammer begins from what was said seventy years earlier, also in an after-dinner address, at a banquet in honor of Max Planck, held in Berlin in 1918. The speaker was Albert Einstein.

> The greatest task of the physicist is the search for those general elementary laws, out of which by pure deduction his picture of the world is formed.

The question he then raises is:

> Has physics, during these seventy years, made some real progress towards achieving what Einstein defined as the greatest task of the physicist? Has physics really reached those general fundamental laws from which it can form a coherent picture of the world?

He goes on (my italics):

> Although physicists, from Los Angeles over London to Leningrad, agree on how to use this theory, they disagree profoundly over what it means. More than fifty years ago, Niels Bohr made the remark, "If anyone says he can think about quantum problems without getting giddy, that only shows that he has not understood the first thing about them." The controversy about the correct interpretation of quantum mechanics is still going on undoubtedly among physicists and philosophers alike. There is still today, about fifty years after the birth of the theory, no scientific consensus on what the mathematical calculus of the theory really describes. The reason for this first debate is undoubtedly the fact that we

face in it one of the greatest conceptual inventions, a revolution largely unnoticed by the general public, not because its implications are not of general interest, but because they are so drastic and exciting that even the scientific revolutionaries themselves hesitate to accept them.

Thus, for example, Einstein, who discovered the quantization of electromagnetic radiation and made a number of other important contributions to the development of quantum mechanics, declared as late as 1949 that "physics is an attempt to grasp reality as it is conceived independently of its being observed." *Well, the generally accepted interpretation of quantum mechanics denies the very existence of such a real world, independent of our perception.* For it claims that quantum entities, such as atoms, electrons, or photons, do not possess attributes of their own, but acquire them only in the act of observation.

In reply to Einstein's objection that it is hard to believe "that," as Einstein put it, "a mouse could drastically change the Universe by merely looking at it," Heisenberg retorted "Atoms are not things" and compared "thing-nostalgic" physicists to believers in a flat Earth. Said Heisenberg, "The hope that new experiments will lead us back to objective events in time and space is about as well founded as the hope of discovering the end of the world in the unexplored regions of the Antarctic." The development of physics in the past few decades seems to indicate that Heisenberg was right and not Einstein. Laboratory experiments performed in the last few years, such as the famous Clauser-Horn experiments in California or the Aspect experiment in France, have demonstrated that atoms and subatomic particles, which people usually envisage as microscopic *things*, fail to have a well defined independent existence or a separate, so to say, personal identity [...] *Many physicists, though not all of them, conclude from such experimental and theoretical results that reality, inasmuch as it has any meaning at all, is not a property of the external world on its own but is intimately bound up with man's perception of the world: his presence as a conscious observer. This conclusion, if justified, is one of the reasons why the quantum revolution is so shattering. All previous revolutions in science, that of Copernicus, or of Darwin, have demoted mankind from the center of creation, to the role of a mere spectator of the cosmic drama. The quantum revolution reinstates man to the center of the stage. Or as Bohr once said, "We are not merely spectators, we are actors in this great drama of nature."* Some prominent physicists, like John Archibald Wheeler for instance, have even gone so far as to claim that the decisive process in the establishment of reality is in the entry of information into the consciousness of the observer.

Taken to its extreme, *this implies that the concrete existence of the world depends on man's perception. We can no longer maintain the idea of a world that exists "out there," independent of us;* instead we have to conceive of it as what Wheeler calls "a participatory universe." For, no elementary phenomenon is a phenomenon until it is an observed phenomenon. In fact, proponents of this view contend that man participates in the creation of the universe [...]

Whether unconventional ideas like these will be accepted or not, the majority of leading physicists are convinced that modern physics can no longer view the world as that physical reality which Einstein had in mind when he made his statement in 1918.

I wonder how Einstein would react were he alive today.

In his autobiographical notes, written in 1949, Einstein discussed, in connection with the Einstein-Podolksky-Rosen paradox, the possibility of rejecting the so-called separability or locality principle, referred to such, for him unacceptable, ideas and called them "spooky" or "telepathic," meaning thereby, of course, that they are not conducive to forming a coherent picture of the world. In other words, we would have to renounce the age-old ideal of really understanding the world we live in. Hawking, with his conception of computers taking over the job of physicists, is not so far from Einstein's position. We would have a mathematically consistent theory of the universe, of its past and its present, but we would not have a coherent understanding of what this mathematics really means.

We may summarize this development, I believe, in simple words: *It seems that the more we know in physics, the less we understand the physical world.* It seems that there is an inverse reciprocality, some kind of uncertainty relation *à la* Heisenberg, between knowledge and understanding. *Surely, the notion of "understanding something" requires a much deeper analysis.*

This means, of course, that physicists, in spite of their spectacular achievements in theory and practice, penetrating into the deepest layers of the atom, conquering the surface of the moon, and exploring the structure of galaxies billions of light-years away, should not be bloated with pride, but remain *modest and humble.*

As an ancient Hebrew passage in the Talmud (Taanith 7a) puts it:

> As waters from the heights descend, that they
> a lower bed may find,
> So, too, with him alone will knowledge stay
> Who hath a humble mind.

5 Aspects

"My suggestion is that at each stage the proper order of operation of the mind requires an overall grasp of what is generally known, not only in formal, logical, mathematical terms, but also intuitively, in images, feelings, poetic usage of language, etcetera" *David Bohm*

OPPOSITION TO ATOMISM

First we return to Newton's experiment, by which he showed how white light can be split with the aid of a prism into a spectrum in which all the colors of the rainbow appear. In Chapter 1, we let Newton himself say how he had carried out the experiment. We saw him sitting there, armed with his prism, in a darkened room into which a single ray of white sunlight was admitted by a small gap in the curtains. If ever there was an apt illustration of Casimir's remark that "scientific theories are a description of a limited number of physical phenomena which in turn make up only a limited part of our human experience," then it is here. From the full, radiant reality of this day in 1666, the researcher let in a ray of light and broke it into pieces with his prism.

We also saw that the experiment and its explanation by Newton came up against a good deal of opposition among prominent contemporaries during the first decades after it. There was not much of this opposition left around 1700. Then about a century later, in 1810, someone again emerged who disputed Newton's theory of color with fire and sword: Johann Wolfgang Goethe.

"What?" you may ask, "Goethe, the poet? The man whose *Faust* was presented in our schooldays as one of the masterpieces of world literature? Was he then an amateur scientist in his spare time?" Yes, indeed it was the same Goethe. What now follows is largely taken from an article by the scientist and novelist Willem F.Hermans, entitled "The Color Theologian," a description which expresses the supreme form of disapproval which this author can think of - and that says something.

I said that Goethe challenged Newton's theory "with fire and sword," and I meant every word. The fierceness, indeed the fanaticism, of Goethe's opposition was out of all proportion. While he did not call Newton a swindler, he did say that he was the victim of a degree of self-deception "which often comes close to dishonesty." He even exhorted students to leave the lecture room when Newton's theory of colors was being discussed. He himself wished that the authorities would ban the theory officially. Goethe as an early incarnation of the Ayatollah!

It goes without saying that Goethe had his own theory of colors which he thought superior to that of Newton. "An amateur scientist?" I asked earlier, but Goethe himself thought otherwise. He regarded his theory as his principal life's work. Anyone could write poems and compose tragedies, but he regarded his theory of colors as a unique achievement. Perhaps we should see his last words, which were already disclosed to me in a German textbook we used at school, in this light: "The last words of the great poet Goethe were, 'Light, more light.'"

We need not spend long on Goethe's theory of colors. "Theology," Willem Hermans calls it, following the German author Albrecht Schöne, who wrote a book on the way in which Goethe dealt with the matter. The theory has left no traces in physics. It has been set aside in a museum of curiosities. We could simply note this and go on with the order of the day. But if we did that, I should not have touched on the whole subject. For the fascinating question which then remains is of course how so enlightened, broadminded, and balanced a person (to use Hermans' words) could make such a mistake. Hermans puts us on the track of finding an answer to this question:

> Goethe's main objection to Newton and similar scientific researchers was not rationalistic, not logical. It was

a feeling. It was purely ethical.

In essence, he thought that what Newton had done was a disfigurement of nature, a violation of holy light. He never tired of crying shame over the *"torture chamber" in which Newton had abused light and broken it into "pieces,"* as an executioner quarters his victim.

So, Goethe's passionate protest was directed against breaking into pieces the "holy light" that formed a whole before the intervention. He called this "torturing." What is broken into pieces cries out with pain. It is a protest against what is called the *atomistic*, and also *reductionist* approach of the scientists. Scientists are splitters. They are the dissectionists *par excellence*. They split white light into colors, and reduce its warm identity to a *figure* (the wavelength). They break the atom into a nucleus and electrons, the nucleus into nuclear particles, the nuclear particles into quarks, and so on. We already encountered earlier the surrealistic machines which produce the gross force by which modern splitting is brought about. In the meantime it has produced the atom bomb. Goethe's protest, stripped of its extravagant form, expresses the feeling that science is on the wrong road if it uses exclusively the atomistic approach. As he said, we cannot be deprived of the right to *wonder at color in all its aspects and significances, to love it and, where possible, to investigate it without torture.* In this formulation we can recognize the language of the poet. Goethe's theory of color could be seen as a first, stunted, probably unsuccessful (though...? see below) attempt at a *holistic* approach by science. That idea is slowly beginning to gain ground in our time.

The atomistic approach is so dominant in physics that it is in fact the only one. It has also unmistakably led to impressive results. A picture of the whole is being built up, starting from the smallest particles, from analysis to synthesis. But the question is whether the picture that we get in this way is not a mutilated one, just as pathological anatomists in the dissection room cannot work backwards and reconstruct a person from a body that they have cut into countless pieces, but at best a corpse.

Before we continue the argument, I want to repeat a story told by James Gleick in his fascinating book *Chaos*. Chaos is a brand-new development in science in the last twenty years; there isn't room to go into it here. However, I do want to mention one aspect of it: the fact that the theory has uncovered a new form of uncertainty in the forecasting of physical events which is totally different from Heisenberg's uncertainty relations, but just as far-reaching. One of the brilliant pioneers in this area is Mitchell Feigenbaum. Gleick relates how he came into contact with Goethe's theory of colors, studied it thoroughly, and came to the conclusion that it was correct. Here are a few lines from his account:

> Where Newton was reductionist, Goethe was holistic. Newton broke light apart and found the most basic physical explanation for color, Goethe walked through flower gardens and studied paintings, looking for a grand, all-encompassing explanation. Newton made his theory of color fit a mathematical scheme for all of physics. Goethe, fortunately or unfortunately, abhorred mathematics.
>
> Feigenbaum persuaded himself that Goethe had been right about color. Goethe's ideas resemble a facile notion, popular among psychologists, that makes a distinction between hard physical reality and the variable subjective perception of it. The colors we perceive vary from time to time and from person to person - that much is easy to say. But as Feigenbaum understood them, Goethe's ideas had more true science in them. They were hard and empirical. Over and over again, Goethe emphasized the repeatability of his experiments. It was the perception of color, to Goethe, that was universal and objective.

That is Gleick's amazing story about Feigenbaum. I reproduce it without drawing any conclusions. I cannot make much of Goethe's theory either, but who knows, that may be because I am also still too molded by the atomistic approach to science. The mere fact that someone of the class of a Feigenbaum sees something in it must at the least be cause for reflection. Perhaps the theory is indeed more than a theology, and finally, after two centuries, Goethe will perhaps get the recognition that he himself had expected that he would get at the latest fifty years after his death.

Back to atomism. In our time the protest against the splitting which is all the rage is beginning to sound louder here and there. I shall go on to demonstrate this with a passage from the book *Things Done* by Jan Hendrik Van den Berg, Professor Emeritus in Neurology at Leyden University: a medical doctor, but also a scholar, one of the very rare people who can look beyond the bounds of their own professional specialism in an attempt to develop a view of the whole person. He is a *homo universalis*, a universal man, and in that he is like Goethe. His total view is set down in a model of human history that he calls *Metabletica*, the title of his best-known book. He writes as easily about psychology, biology, and mathematics as about architecture, the graphic arts, and music. He too is a loner, who is open to much criticism from specialists on the subjects into whose territory he ventures, because of course they know the details better, but his view is extremely attractive and for me is even more than that..

The passage I mentioned is about one of the most famous splittings in the history of science, the experiment carried out by Lavoisier in 1783 in which he showed that the "element" water can be split into two others, hydrogen and oxygen. I put "element" in quotation marks because until the moment when Lavoisier did his experiment, it had been thought that the world in which we live is made up of four elements - water, air, fire, and earth -, a division which comes from classical antiquity. Up to that point, it had occurred to virtually no one that these elements might be split up further. The term "element" in this old view expresses more than the character of indivisibility. There is something "elemental," something essential about it. "Think of water, for example," Van den Berg suggests:

Take a glass of pure water from a mountain brook, from a spring, drink that water and you have the sense of taking in elemental matter. No oxide. No combination of two gases. *Water* - which is not clear by chance. Air is even clearer. Mountain crystal of a pure kind is clear. Fire makes clear, purifies, is purity in action.

Then follow these words:

What did Lavoisier really demonstrate? That water is not an element. Water is a composite. Water consists of two other materials. Why not say that Lavoisier broke water into pieces? Lavoisier destroyed water.

But that is not all. Water, this element, reveals itself in endless difference. Spring water, river water: Assuming that the water in it is pure, the water from the gushing spring is still not like the water of the slowly flowing river. Rain water, ground water. Drinking water, baptismal water, holy water. The chemist will observe that it is always literally the same thing, with the same formula H_2O. Who can deny that? But now it can also become clear, or at least the supposition can arise, that the designation H_2O betrays water, the manifold varieties of the element water. Drink H_2O - if you can. Go and swim in H_2O - if you want.

From homogeneous, standardized, democratized water a variety of things emerged. The cause of this development lies in the *process of division*, in the principle of division that splits water into two materials, hydrogen and oxygen. Could not perhaps these two materials also be divided? I hope that the reader will consider the question inevitable. Indeed it was raised, and in the meantime the answer has become known. The new elements, hydrogen and oxygen, like all the other elements in Mendeleyev's system, can be divided, split into a number of very small particles [...] When the second, further, splitting, which standardizes the original elements even more (from now on they are also like one another), emerges from the first such splitting by Lavoisier, the direct consequence of the second splitting is the atom bomb of 1945.

Thus Van den Berg. "In that case," one might ask, "is water not H_2O?" No, water is not H_2O. At most it is also H_2O. H_2O is one aspect of water-ness. Anyone who absolutizes the splitting in elements abandons the "elemental" character of water.

"All right," someone might perhaps say, "I get the gist of the protests, but they're not very impressive. A poet and a doctor. A lot of passion and not much science. Like the woman in the 'Delft marriage' who says, 'I can't fault your arguments, but there's more to it than that. I feel that things are different.'" Hence the following quotation:

Nevertheless, this sort of ability of man to separate himself from his environment and to divide and apportion things ultimately led to a wide range of negative and destructive results, because man lost awareness of what he was doing and thus extended the process of division beyond the limits within which it works properly. In essence, the process of division is a way of *thinking about things* that is convenient and useful mainly in the domain of practical, technical and functional activities (e.g., to divide up an area of land into different fields where various crops are to be grown). However, when this mode of thought is applied more broadly to man's notion of himself and the whole world in which he lives (i.e. to his self-world view), then man ceases to regard the resulting divisions as merely useful or convenient and begins to see and experience himself and his world as actually constituted of separately existent fragments. Being guided by a fragmentary self-world view, man then acts in such a way as to try to break himself and the world up, so that all seems to correspond to his way of thinking. Man thus obtains an apparent proof of the correctness of his fragmentary self-world view though, of course, he overlooks the fact that it is he himself, acting according to his mode of thought, who has brought about the fragmentation that now seems to have an autonomous existence, independent of his will and his desire.

The quotation comes from the book *Wholeness and the Implicate Order* by the contemporary English physicist David Bohm, whom we met earlier. We shall now investigate his approach more closely.

DAVID BOHM AND UNDIVIDED WHOLENESS

Holism, the view that the reality in which we live is one whole and that everything is connected with everything else, is "in." A possible link between this view and the world of modern science has been popularized above all by the books of Fritjof Capra, *The Tao of Physics*, and Gary Zukav, *The Dancing Wu-Li Masters*, in which the authors make connections between quantum physics and the world of the ideas of the Eastern religions. Their insights are disputed, and have generally been received coldly in the world of scientists. The only writer in this context who is taken seriously in the world of physics is David Bohm, above all, I think, because he is a prominent physicist and because he does not stop at stories and philosophies, but makes serious attempts to translate them into terms of authentic physics. He has set out his ideas for a wider public in the book I mentioned earlier. I shall try to give a summary of it here.

First of all, Bohm discusses the significance of the concept of "theory." The root of the word theory is a Greek word which means "observe." A theory is a particular way of looking at an object. A theory is comparable with a particular *view* of an object. Any view from a particular direction shows us a particular *aspect* of the object. The whole object is never perceived in one view but is only experienced *implicitly* as the one reality which shows itself in all these perceptions. All our different ways of thinking must be regarded as different ways of looking at the same reality, each of which is clear and applicable in a particular sphere. Theories are not direct descriptions of "reality as it is."

That also goes for atomic theory, which was first formulated more than two thousand years ago by Democritus. This has certainly done good service in achieving a particular insight into particular aspects of reality, but on the whole the theory is no longer seen as an insight or a way of looking, but as an absolute truth about reality. People began to think that the whole of reality in fact consists only of atomic "building blocks" which influence each other in a more or less mechanical way. Atomic theory thus developed into the most important buttress to a divided approach to reality. This perspective has also had an effect on other sciences, like biology, where people are often convinced more strongly than in physics of the correctness of this idea, above all because there is no awareness (or hardly any awareness) of the revolutionary character of developments in modern physics. Most molecular biologists, for example, believe that the whole of life and the human mind can ultimately be understood if only enough research is undertaken into the structure and function of DNA molecules. This kind of idea is also beginning to dominate psychology. The remarkable consequence is that faith in the divided, atomistic approach to reality is strongest precisely in the sciences which study life and the mind and in which the "formative cause" is most clearly perceptible in unbroken and flowing movement.

"All right," someone might say, "but we see this atomistic structure, and is that not a principle?" Bohm's answer is that anyone who approaches nature with atomistic blinders gets reactions to fit the approach. If we confront fellow human beings with the theory that they are enemies against whom we must defend ourselves, they will react as enemies and so "confirm" our theory. So, too, nature reacts in accord with the theory by which it is approached. A predominantly atomistic approach will hardly seek evidence to the contrary, and if this evidence is nevertheless found, as in modern science, it will be thought to be of little interest or completely denied.

Bohm finds this evidence to the contrary specifically in quantum physics and the theory of relativity. That these two are irreconcilable is a cloud that has long hung over modern science, a cloud which in Bohm's view has already taken on the dimension of a thundercloud. Apart from this, quantum physics shows that it is much more logical to see reality as an undivided unity. At an earlier stage, we encountered the two particles in the EinsteinPodolksy-Rosen experiment which were connected with each other in a "telepathic" way. What is more logical than to understand them as an undivided unity? Nor can the difference between the observer and the observed be maintained any longer. These are *aspects* of one whole reality which overflow into each other and penetrate each other, a reality which is indivisible and cannot be analyzed. Bohm also concludes from a consideration of the theory of relativity that there is no place in it for the idea that the world consists of "basic building blocks." Instead of this, the world must be regarded as a universal stream of events and processes. In this totality, the atomistic insight is a simplification, an abstraction which is valid only in a limited context.

Bohm provides illustrations of his view. The old accepted way of describing reality has parallels with the way in which a lens works, for example in a camera. The lens projects a picture of an object. The picture is what we perceive. The image from the lens is one on one: One part of the picture corresponds with each part of the object. The sum of the images of all the parts of the object produces one coherent picture. That is how scientists have set to work hitherto: Reality has been divided into parts which are perceived and depicted separately. In this way, a total picture (synthesis) is constructed from the analysis of the separate parts, a picture of the world (but this is still only a picture). This method of analysis is strongly encouraged by the success of the lens as an instrument of observation in science. When developed into a telescope, it makes it possible to observe fantastically distant stars; in the form of the electron microscope, one can also see atoms with it. This led to the idea that the "lens-like" approach was usable and valid always and in all circumstances.

However, modern science, and above all quantum theory, has shown that an analysis of reality into separate, clearly limited, parts is no longer relevant. Quantum theory, rather, indicates a reality which must be seen as an undivided totality. That can also be illustrated by means of an optical method of representation, *holography* (literally, the writing of the whole). Here a beam of laser light is made to fall obliquely on a semi-transparent mirror. Part of the beam is reflected back and falls on the object that we want to observe. The object disperses the light, part of which falls on a photographic plate, along with the other part of the original beam which was admitted through the mirror. The two beams begin to interact there ("interference"), and the interference pattern is fixed on the plate. It is usually so fine that it cannot be perceived with the naked eye.

If we now let a beam of laser light fall on the plate and keep our eyes behind the plate, we see the whole object which was originally illuminated in three dimensions. *However, if we illuminate only a small part of the plate, we still see the whole object, but rather less sharply.* So there is no question here of a one-on-one image, in which each part of the object is fixed on one part of the photographic plate. *On the contrary, each part of the "image" on the plate has a connection with the whole structure that is imaged.*

In this approach, the photographic plate is not of essential importance. It merely fixes the interference pattern which is present at this point in space. Even if the plate is not there, the light in any part of space contains the information about the whole object. The information is *implicitly* present (enfolded there). In Bohm's language, "every area in space and time implicitly contains a total order." In the carrying out of an experiment, aspects of this total order can come to light, be made *explicit* (unfolded).

This is made clearer by a second illustration. Take a transparent container filled with a viscous fluid, for example glycerine. The container is equipped with a mixing apparatus which can mix the fluid slowly and regularly. We now put a drop of insoluble ink into the fluid and turn on the mixing apparatus. The drop then changes gradually into a thread which becomes increasingly thin and spreads over the whole fluid. In due course, the thread as such can no longer be seen and the fluid has taken on a vague inky color. The drop of ink is enfolded, implicitly present, in the fluid. We now put the mixing apparatus into reverse. The whole process now takes place in the opposite direction: The thread rolls up again and at a particular moment the drop again becomes visible. It is "unfolded," made explicit.

We now extend the experiment as follows. After putting in the first drop, we mix a hundred times, put in a second drop close to the place where we had placed the first, mix another hundred times, and so on, until we have added, say, twenty drops. All the drops are now enfolded in the fluid. If we now put the mixing apparatus into reverse and turn it back quickly, then first the last drop to be put in will appear and then be enfolded again, after that the next to the last one, and so on. What we perceive, think we perceive, is one ink drop which moves in a straight line through the fluid. Because it goes so quickly, the line seems to become continuous, but on closer inspection it is a discontinuous, intermittent movement in twenty stages, after which the "particle" has disappeared.

The experiment I have described shows parallels with the way in which traces of "elementary particles" are made visible in a photographic emulsion. In terms of the classical understanding, the particles on their way through the emulsion cause a series of chemical reactions which, after the plate has been developed, become visible as an apparently continuous but in reality discontinuous crooked or straight line, the "path" of the particle; however, in the context of quantum mechanics, in Bohm's language, the trace perceived is never more than an *aspect* which becomes visible in direct observation. It is an aspect of the implicate order that is made explicit as a consequence of our mode of perception. The movement must be described discontinuously, with the help of "quantum jumps," and that means that the whole idea of an accurately defined path of a particle by which the visible points of the trace are connected with each other has become meaningless. The use of the word "particle" in this quantum context is thus extremely misleading.

In Bohm's view, reality is an undivided whole in which the "implicate orders" are borne up by the *holostream* or *holoflux*. In the illustration, the holostream is the moving glycerine fluid in which the "orders" of the ink drops are enfolded. Sometimes particular aspects of the holostream (light, electrons) can be made explicit, but on the whole all aspects of the holostream flow indivisibly into one another. In its totality, the holostream is in no way limited and is not defined by one particular order or measure. So, the holostream is *indescribable and unmeasurable*.

Thus if we attach primary significance to the holostream, it makes no sense to speak of a *fundamental* theory which can be a *permanent* basis for the *whole* of science and from which *all* scientific phenomena can ultimately be derived. *A theory abstracts a particular aspect which is valid only in a limited context.* Thus the construction of a theory is comparable with the work of a painter who paints on canvas a particular aspect of reality as he or she sees it. A theory is *a work of art. Science is an art-form*, one of the art-forms which are concerned with the contemplation of reality. Just as the definitive canvas will never be painted and the definitive poem will never be written, so science will never produce the definitive theory of reality. Thank God, I would say, and if you have read Chapter 3 you will see what I mean. It means that life can go on.

A nice story, you might say, but only a story. How do you substantiate something of that kind? Now Bohm is also busy translating his ideas into ordinary, hard, science. It is difficult for me to judge how far he has got with this. But here perhaps is an indication.

In the spring of 1987, the Netherlands Scientific Association held a symposium devoted to holism. The speakers were two philosophers and a physicist. The physicist, Jan Hilgevoord, paid a good deal of attention to Bohm's formal theory. He ended his argument with the words: "I don't know what our final view of all this will be, but it seems certain that quantum mechanics confronts us with the limitation of our habits of thinking and that we may perhaps have to season them with a pinch of holism." A pinch of holism (whatever that may be) is still not much, but if the scientists, Pais' "conservative revolutionaries," have got that far, a breakthrough might not be far away.

THINGS HAVE THEIR MYSTERY

What is gold? According to scientists, gold is a simple material, an element which in a solid state consists of atoms which are all alike and which are stacked in a regular way in a structure which you also get if you put a collection of ping-pong balls as close together as possible. Each atom consists of a nucleus with a positive charge around which sixty-five electrons circle. In solid gold the outermost electron is no longer strictly attached to the atom to which it originally belonged, but wanders more or less freely through the whole crystal. This "gas" of free electrons is the reason why gold conducts electricity so well. This is a very simplified view of things which could still be modified by a quantum-theoretical approach. From this, there emerges an "electron structure" which we must know if, for example, we are to understand what happens when we cast some white light from outside on a piece of gold. We know what happens: The gold has a yellow color, which is caused by the fact that the electrons absorb part of the light in the blue part of the spectrum. Because we know the electron structure of gold, we can also calculate that. It fits. We understand why gold is yellow.

I could go on like this for many more pages. Gold has a structure and has properties, and the properties can be understood from the structure. That is the task which solid-state physicists have posed themselves, and they have made a good deal of progress in this direction.

The showpieces of the National Museum in Athens include the golden death masks which were found in the royal tombs at Mycenae. The public that shuffles in hordes along the glass cases to see this gold knows little of what I have just said. For them, this gold has a property which does not appear in the series of physical properties (electrical, mechanical, magnetic, optical, etc.): It arouses emotion. It is gold which has been shaped by the hand of the artist into an object that has expression, speaks a language. *It is gold with a soul.* A physicist who examines these masks with his instruments does not see any of this. The material of the masks has the same electrical conductivity and the same coefficient of reflection as that of a gold wrist watch.

The Colombian Gabriel Garcia Marquez, author of such splendid novels as *One Hundred Years of Solitude* and *Love in the Time of Cholera,* and a winner of the Nobel prize for literature, is reckoned to be one of the "magic realists." These are writers who describe the reality around them as though it had a soul. He himself thinks that this is nonsense. Here is a quotation in which he explains why:

> They call magic realism everything that happens in the Caribbean or Latin America, or that is somewhat strange and unusual. In fact there is no magic realism in literature. There is a magical reality which you can find in the Caribbean. For that you only have to go down the street. We have grown up with this magical reality. But it is also present in Europe and Asia.
>
> However, you are hindered by your cultural formation. All Europeans are ultimately *Cartesians.* They reject everything that does not fit rational thought. You go to Europe and you see just as amazing things happening there as in our countries. It's just that we surrender ourselves to reality more easily. We are part of it, we accept it. *Your system of thought compels you to reject reality.* Things have gone very well with you in life. But I believe that you enjoy yourselves less than we do.

So, there is no such thing as magic realism, but only a magic reality. He sees the ordinary and writes down what he sees. So, he is just a realist. We in the West fail to see it because we close ourselves to it, because we have "Cartesian" blinders, because we look on the world with the rational eyes of scientists, the eyes of the "Enlightenment," which in this context is an amusingly ironic word.

What is a sunflower? To discover that you could consult a botanical book of flowers and there find the following description:

> *Helianthus* (sunflower): A large annual herb with a stout erect stem, often unbranched, leaves alternate, petiolate, broadly ovate. Three-veined sinuate toothed, hispid with stiff hairs above and beneath. Disk-florets brownish, ray-florets golden yellow.

The painter Vincent Van Gogh brought the sunflower to life in various of his paintings. Let's listen to what the neurologist Jan Hendrik Van den Berg says about them.

> Anyone who stands in front of Van Gogh's sunflowers is struck by the fact that the blooms are both very like and yet not like the sunflowers we know. They are not like the sunflowers that we think (or thought) we know, because no photograph can capture such sunflowers on a color film. They seem particularly like the sunflowers that we increasingly begin to see because they contain and reproduce all that there is about a sunflower. The tall growth of the plant with its powerful flower. The yellow flames of the petals branching out from the head. The bulbous base and increasing density of the seed head. The heat of a sun-drenched summer. The majesty of the sun itself. The irresistible splendor of the earth. It is all far more than the flower itself. But that is not true. The "more" belongs to the flower, is this flower. Take away from the flower all that I have just mentioned, and one has a botanical plant, a poor remnant of what the sunflower is. In the two centuries between 1700 and 1900 we learned, step by step, to be content with the poor remnant and thus came to think that the qualities I have summed up are *ideas* attached to the sunflower. Van Gogh unmasked this as falsifica-

tion. He demonstrated that the ideas are not ornaments, but authentic properties of the sunflower, *of the thing itself*. His sunflowers show that the world in which we live is a charged, numinous, magical world. The thing is space, time, color, flower, noise, silence, passion, faintheartedness. We encounter a creation in the *thing*. That is, if we are willing to do away with that eye, that ear, those senses trained between 1700 and 1900. We have to be willing to do away with the thing between 1700 and 1900 and want to see the thing of the twentieth century. Van Gogh did that for us.

Van den Berg calls that "properties." Majesty, splendor, these are things that will not be indicated on any measuring instrument used by physicists. Yellow flames, the temperature of which cannot be measured. Anyone who wants to see this must remove the blinders of the period between 1700 and 1900, the period of the Enlightenment. Is it then surprising that Van Gogh's contemporaries didn't see it? Isn't it splendid that anyone who wants to see an exhibition of his work nowadays must buy a ticket long beforehand? The period between 1700 and 1900 is at last coming to an end, not least because of the achievements of twentieth-century science.

ASPECTS OF LIGHT

There is perhaps no natural phenomenon which has so fascinated physicists, and not only physicists, so much as light. In the meantime, we have come to know a great deal about it. Light, like all other electromagnetic waves, moves with the improbable velocity of 186,000 miles a second. In one second it could go round the earth at the equator eight times. (Just the experiment with which Foucault was able to measure this fantastic velocity for the first time should convince anyone that science is a splendid subject.) There is another amazing thing about the speed of light. If you shoot a bullet in the direction of travel from the roof of a train going at

60 mph and it leaves the barrel of the gun with a velocity of 30 mph, then to an observer on the ground the bullet has a velocity of 60 + 30 = 90 mph. If you shoot in the other direction then that becomes 60 - 30 = 30 mph. The two velocities of the bullet can be added. If you put a source of light on the roof (a machine gun which shoots photons), then for an observer on the ground, the velocity of this light is the same in all directions: 186,000 miles a second. The speed of light cannot be influenced. It is at the same time the greatest speed which is possible in nature. This datum, accepted by Einstein as a hard fact, formed the basis of his special theory of relativity, from which amazing consequences derive for the concepts of space and time, and which at the same time produced the explosive formula $E = mc^2$.

Sometimes light shows us its "particle aspect": It behaves like a stream of pellets from a shotgun, a stream of photons which can collide with other particles, like electrons, in the same way as two billiard balls. Sometimes it shows us its "wave aspect." White sunlight then appears to consist of waves with a wide range of wavelengths, of which we can see only a small part: The human eye is sensitive only to the range of wavelengths between 0.4 and 0.8 thousandths of a millimeter. In between, we perceive the spectrum of colors; outside we feel the infra-red radiation of heat, and the ultra-violet part of the spectrum has an effect on our skin, giving people that sought-after tan.

All these are examples of "hard facts" about light, information that light has given up to "observers," people who have studied light in order to penetrate its secrets. As far as I am concerned, they are so many wonders at which one never ceases to be amazed, experiences of people who look at the light.

Scientists are not the only ones who experience the light, and who set down their experiences in a work of art: their theory, their way of looking at things. Artists (other artists!) also do that. Painters, of course. Wasn't Rembrandt called "the magician with light"? And poets. The following poem by the Dutch poet Hans Andreus relates an experience with the light.

Lying in the sun

I hear the light the sunlight pizzicato
the warmth is again whispering to my face
I lie again its like that but not like that
I lie a monomaniac and monofool with light

I lie outstretched lie singing in my skin
lie singing gently answering the light
lie foolish not so foolish outside human beings
singing things of the light that lies on and around me

I lie here brotherly very southerly lie without
knowing how or what I am only lying still
I only know the light of miracle upon miracle
I only know all that I want to know.

A wonderful experience with wonderful light. Light that
emits a strumming noise that you can hear, to which you
can sing an answer. Light with which you have an understanding. Light that does all the things that physical light
cannot do. To the poet's "theory" this is no problem. He
does not know "how or what." He knows only all he wants
to know.

In the following poem by Ida Gerhardt, which for me is
one of the two or three best in the Dutch language, the light
has the first and last word.

Sunday morning

The light starts walking through the house
and touches things. We eat
our early bread, baptized in sun.
You have spread the white cloth
and put grasses in a vase.
This is the day on which work rests.
The palm is open to the light.

What a rest! The light does not rush into the house at a velocity of 186,000 miles per second; it *walks*. As it walks, it touches things gently and wakes them to life. If the light in the first line is already no longer physical light, that in the last line is explicitly metaphysical, religious by nature. It is other light. Through the week, the palms are turned downwards; one sees only the backs of the hands of busy people. On Sunday morning, they open the palms of their hands, the gesture of someone calling for help from outside, who turns to the light, to God. It is the gesture of the Christian community, who, meeting on Sunday morning, hear as the first words spoken by the minister, "Our help is in the name of the Lord who made heaven and earth," and who at the end of the service again open the palms of their hands to the light as they receive the blessing, "The Lord bless you and keep you. The Lord lift up the light of his countenance upon you and give you peace."

One could argue that the light as it is spoken of in poetic or religious language is "only" a metaphor and that the only "real" light is that which physicists talk about, objective light. Is that the case? Is God an obsolete notion which has been made superfluous by the process of science? Or is God, as some contemporary scientists claim, to be found precisely through science? Or are there perhaps other ways which lead to God? Can anything be "proved" in this area? Those are questions which we must go on to consider.

6 Four Physicists and God

This book is not, or not primarily, about science in the abstract sense, but about people who do science, about their views of human beings and their views of God. So I want to open the discussion of the question of God which I indicated with a chapter in which we shall investigate what role God plays in the life and world-view of four prominent physicists. Two of them, Newton and Pascal, lived before 1700; the two others, Einstein and Hawking, after 1900. This gives us four very different pictures, but they make it clear that there is a gap between the first two and the last two.

ISAAC NEWTON: A MAN ALL OF A PIECE

> "One and the same I am throughout life in all the organs of the senses: and one and the same is God, always and everywhere."

If you distributed ballot papers to a number of scientists and asked them to put down the three greatest physicists in history in order of importance, the result would show a number of different names in third place. Galileo, Maxwell, Bohr, and Heisenberg would certainly be found there. But the two names in the first two places would be the same on all the forms: Newton and Einstein, not necessarily in that order.

We have already met Newton on a variety of occasions: as the developer of a new theory of light, set down in his book *Opticks* of 1704, and, even more important, as the founder of a new mechanics and the discoverer of the general law of gravity, first published in 1687 in his book *Philosophiae Naturalis Principia Mathematica*, usually referred to as the *Principia* for short. That book changed not only science but also the world. We won't go into that further here. This chapter is about something else.

All the biographies of Newton (1642-1727) say that as well as science he also did theology. The chapter on "Newton's Theology" in these biographies is brief compared with the rest and with good reason. His academic publications in this area, in contrast to his scientific work, have not stood the test of time. You will not come upon references to Newton in any modern theological works.

To answer questions like "What kind of a man was Newton?", "What were his deepest motives?", "Who was God for him?", we need more than can be gathered from his theological works. His posthumous writings contain a good deal of information about those questions. They are extraordinarily extensive. Zealous statisticians have established that Newton put down on paper about a million words on religion, about as much as on science. Most of them were not published. Certainly it is striking that in his great scientific works, too, here and there he speaks frankly about God, just as we noted earlier that the man Newton could also be found in them. Recently a book written by Frank E.Manuel called *The Religion of Isaac Newton*, largely based on Newton's unpublished writings, has made an investigation of his religious life. In particular the Yahuda Collection (Jewish National and University Library in Jerusalem) plays an important role in it. I suspect that what Manuel has brought to light from his wealth of documentation is far from generally known and that it is worth summing up his findings briefly here.

One extremely important correction to the current picture must be made immediately. To think that "Newton's religious life" means that he had other lives alongside this, for example a scientific life, is to have a wrong idea of things. But this is the image that one usually encounters in the literature and which was expressed in its most extreme form for the first time by the French astronomer Jean-Baptiste Bidot at the beginning of the nineteenth century. This introduces a division into Newton's life: first the period of youth and young adulthood, to about the age of forty, during which he carried out experiments and developed theories according to the strictest possible scientific method, the time of his revolutionary discoveries. Then came the second period, in which he lapsed into senility and gave himself over to mystical fantasies.

Even the most superficial investigation, says Manuel, shows that this conception of things does not have a leg to stand on. From Newton's earliest writings, dating from about his twentieth year, his science and religion are both present, to the same extent and intensity, and inextricably connected. They form a unity. His whole life can be summed up in three words, "servant of God." God is his Lord, his Master, and the search to do God's will and obey it was all he was concerned with. In looking for knowledge of his Master and his will only two ways were open to him: studying God's actions in the physical world, his creation, the book of nature, and studying God's commands in the book of scriptures. For Newton, the two books were of equal worth, and were not in tension with each other. The knowledge of God disclosed in the one is in harmony with the other; they form a unity. In both, Newton finds the same characteristics of the work of the Creator: a divine simplicity in nature and scripture.

Newton's God is above all a personal God. This has perhaps been most attractively described in the third edition of the *Principia*. Here it is in the English translation (from the Latin) of his pupil William Whiston:

> This Being governs all Things, not as a Soul of the World, but as Lord of the Universe; and upon Account of his dominion he is stiled Lord God, supreme over all. For the word God is a relative Term and has Reference to Servants and Deity is the dominion of God; not (such as a Soul has) over a Body of his own, which is the Notion of those who make God the Soul of the World; but (such as a Governor has) over Servants. The supreme God is an eternal, infinite, absolutely perfect Being; but a Being, how perfect soever, without Dominion is not Lord God. For we say: my God, your God, the God of

Israel, the God of Gods and the Lord of Lords. But we do not say: my Eternal, your Eternal, the Eternal of Israel, the Eternal of the Gods. We do not say: my Infinite (your Infinite, the Infinite of Israel). We do not say: my Perfect (your Perfect, the Perfect of Israel). For these Terms have no Relation to Servants.

For Newton, that was the crux of the matter, to which he held with great tenacity, as in the disputes with his contemporary Leibniz, whose theology was in Newton's eyes a rationalistic structure and whose God was a "metaphysical" God. Newton's God is a personal Master, to whom the servant says, "My Lord." Concepts like love and grace stand in the background for him, but they are not absent.

This personal relationship to God is also tangibly present in a moving document from 1662 (when Newton was twenty). It is a written confession of guilt of the kind that was often practiced at that time: Before taking part in the eucharist, by way of preparation believers set down a number of their sins on paper. Among a number of trivial sins, like small disobediences and lies, we read there:

Not turning nearer to Thee for my affections. Not living according to my belief. Not loving Thee for Thyself. Not loving Thee for Thy goodness to us. Not desiring Thy ordinances. Not longing for Thee. Fearing man above Thee.

Newton's view of the way in which theology had to be done in relation to the church and believers is worth mentioning, and if it had gained acceptance, it would have saved the Christian church a lot of misery. From his perspective, it was enough for the great majority of "ordinary" believers to subscribe to the "primitive" apostolic creed and live in accordance with the commandments. They might even differ over the precise meaning of the words used in it. The practice of theology was reserved for a select company, among whom he counted himself. This study was of great importance, but not absolutely necessary. At this level there might, indeed must, be fierceness, but differences of opinion were not to have any consequences in the form of persecution or exclusion from the Christian community. He thus argues for "free space." Even for the select company, subscription to the basic principles must be enough.

Newton's personal "primitive" creed can be found in his book on church history. I quote it here not because it is so original but because of its simplicity and authenticity.

> We must believe that there is *one God* or supreme Monarch, that we may fear and obey him and keep his laws and give him honour and glory. We must believe that he is the father of whom are all things, and that he loves his people as his children that they may mutually love him and obey him as their father. We must believe that he is *pantokrator* Lord of all things with an irresistible and boundless power and dominion that we may not hope to escape if we rebell and set up other Gods or transgress the laws of his monarchy, and that we may expect great rewards if we do his will. We must believe that he is the God of the Jews who created the heaven and earth and all things therein as is expressed in the ten commandments that we may thank him for our being and for all the blessings of this life, and forbear to take his name in vain or worship images or other Gods. We are not forbidden to give the name of gods to Angels and Kings, but we are forbidden to have them as Gods in our worship. For though there be that are called god whether in heaven or in earth (as there are Gods many and Lords many) yet to us there is but one God the father of whom are all things and we in him and one Lord Jesus Christ by whom are all things and we by him: that is, but one God and one Lord in our worship.

After Newton's death the period of the Enlightenment dawned, in which Reason was elevated to the status of a deity. In 1802 Henri de Saint-Simon, a French ex-aristocrat, incited his contemporaries to found a new church in which the scientists would function as priests; he called it the Religion of Newton. At the same time another French aristocrat issued a manifesto in which the English people were roundly condemned for their failure to revere Newton's divine person: He began a new chronology with the year of Newton's birth and proposed the foundation of a sanctuary in Woolthorpe, the village in which Newton was born. Had Newton heard of this, he would have turned in his grave. That grave is in Westminster Abbey. It bears the following inscription:

HERE LIES
SIR ISAAC NEWTON, KNIGHT,
Who, by a vigour of mind almost supernatural,
First demonstrated
The motions and Figures of the Planets,
The Paths of the Comets, and the Tides of the Ocean.
He diligently investigated
The different refrangibilities of the Rays of Light
And the properties of the Colours to which they give rise.
An Assiduous, Sagacious, and Faithful Interpreter
Of Nature, Antiquity, and the Holy Scriptures,
He asserted in his Philosophy the Majesty of God,
And exhibited in his Conduct the simplicity of the Gospel.
Let Mortals rejoice
That there has existed such and so great
AN ORNAMENT OF THE HUMAN RACE
Born 25th Dec.1642, Died 20th March 1727.

Perhaps a bit exaggerated, but it is not far from the truth. Stephen Hawking, the present occupant of the chair once held by Newton, devotes two pages to Newton in his book *A Brief History of Time*. The first sentence of his account reads, "Isaac Newton was not a pleasant man." And the last, "He successfully conducted a major campaign against counterfeiting, even sending several men to their death on the gallows." In between them, without one positive word, there is the same heavy-handed approach. What drives someone of Hawking's position and Hawking's class to such negativity escapes me completely, but it is a good reason for ending with a comment that puts everything in perspective. I prefer to do so with the words of Frank Manuel:

> Newton showed himself to be a master of academic debating techniques and he would demolish an enemy on his own ground, with his own weapons and with a certain cruel satisfaction. I am not recommending Newton for sainthood.

BLAISE PASCAL: TENSION

> "The heart has its reasons, which reason does not know"

Quantities in physics are expressed in units which often bear the name of a scientist who played a major role in this area. So electrical units of current, potential, and power are expressed in terms of Ampère, Volt, and Watt, names known by any owner of an electric light bulb. The unit of *force* is the Newton, and the unit of *tension* is the Pascal.

That phenomenon also has a symbolic significance for me. Anyone who is confronted with Newton encounters force. He was a strong personality, a man all of a piece, a monument. "A vigour of mind almost supernatural" is written on his tomb. He was impressive, but remote. Anyone who encounters Pascal enters a field of tension, the tension between heart and understanding, between the "order of the mind" and the "order of love," as he puts it; between human beings in their greatness and their deep fall; between God who is hidden and yet near. Here is the tension between the immense task with which he saw himself confronted and the short time that was given to him. We encounter a *person* who is still close today.

First a brief account of his life and work.

Pascal (1623-1662) never went to school; he was taught by his father, who discovered very quickly what an extraordinarily gifted pupil he had, an infant genius. Already at the age of sixteen he enriched mathematics with a new theorem. Before he was twenty, he constructed the first computer in history. He was one of the first to give priority in science to proof from experiments and to work according to a strict scientific method. He gained his greatest fame from the experiment in which he showed that the level of the column of mercury in a barometer decreases when one moves the barometer from the bottom of a mountain to the top, thus demonstrating that the mercury is kept up by the weight of the air. "Pascal's law" also has permanent significance in the science of the equilibrium of liquids. Pascal combined a brilliant understanding, a lively imagination, and an acute critical capacity with a boundless urge to work. The years which he devoted to making his computer, his sister Gilberte wrote in the first biography of him, ruined his health permanently. From his eighteenth year he did not have a single day without pain. He knew that he had little time to fulfil what he saw as his task. He died at the age of thirty nine.

Pascal grew up in a devout Catholic family. Of decisive significance for his religious development was the fact that when he was twenty-three, in 1646, the family came into contact with the spiritual renewal movement within the Catholic church which had its center in the monastery of Port-Royal. The movement supported a return to belief in the basic values of Christian faith and therefore soon came into conflict with the official Catholic church, in which this experience had fallen into the background and had been overgrown with numerous abuses. Pascal was the first of the family to be completely won over by Jansenist insights, as was later, through him, his sister Jacqueline (to whom he was very close all his life; she later entered the convent of Port-Royal), and subsequently the whole family. Pascal played an important part in the dispute between the movement and, in particular, the Jesuits, to which his apologetic work, the *Lettres Provinciales*, bears witness. The letters were condemned by Rome. As a result of this, Pascal later called the Inquisition and the Order of Jesuits "the two scourges of the truth."

The aim of the movement was, as Gilberte describes it, to live for God and to have no other goal than God alone. "This truth," she goes on, "seemed so obvious to him [Pascal], so necessary and so useful, that he ended all his researches, so that from then on he renounced all other knowledge to devote himself to what Jesus Christ called the one thing necessary." It is almost ironical to note that, as in the case of Newton, an attempt was made to divide the life of Pascal into two, but with precisely the opposite aim: Gilberte regards the first part of Pascal's life, in which he was occupied with science, as a kind of youthful sin, to which he put an end when he saw the light. As in the case of Newton, this view of things does not take account of the facts. All his life Pascal was occupied with science and also with religion, although the accents indeed shifted over the years. There was no question of a conflict between the two.

Perhaps the most important thing that Pascal has left to us is his *Pensées*. After his death, around a thousand loose sheets were found on which in a very pointed way he had written his thoughts about science, human beings and their relationship to God, and about Christian faith. They were later deciphered, arranged according to subject matter, and published. They are still as fresh as ever. Simply because of the nature of the *Pensées*, it is impossible to give a summary account of them. In what follows I shall try to indicate some lines in Pascal's thoughts on religion.

The person we encounter in Pascal is someone who, face to face with eternity and the immense space of the universe, is overwhelmed by a feeling of nothingness and who desperately asks where his place in it is and what the meaning of his existence can be:

When I consider the short duration of my life, swallowed up in the eternity before and after, the little space which I fill, and even can see, engulfed in the infinite immensity of spaces of which I am ignorant, and which know me not, I am frightened, and am astonished at being here rather than there; for there is no reason why here rather than there, why now rather than then. Who has put me here? By whose order and direction have this place and time been allotted to me?

There is a great temptation, confronted with such questions, to give in quickly, to lapse into skepticism, and to take refuge in an existence that is filled with diversion in "pleasing objects":

> When I see the blindness and the wretchedness of man, when I regard the whole silent universe, and man without light, left to himself and, as it were, lost in this corner of the universe, without knowing who has put him there, what he has come to do, what will become of him at death, and incapable of all knowledge, I become terrified, like a man who should be carried in his sleep to a dreadful desert island, and should awake without knowing where he is, and without means of escape. And thereupon these wretched and lost beings, having looked around them and seen some pleasing objects, have given and attached themselves to them. For my part, I have not been able to attach myself to them, and, considering how strongly it appears that there is something else than what I see, I have examined whether this God has not left some sign of himself.

How do you discover whether God has left some sign of himself? Pascal did not find the "proofs of God" which were current in his time and which had been produced by scholastic thought very convincing:

> Nature presents to me nothing which is not matter of doubt and concern. If I saw nothing there which revealed a Divinity, I would come to a negative conclusion; if I saw everywhere the signs of a Creator, I would remain peacefully in faith. But, seeing too much to deny and too little to be sure, I am in a state to be pitied.

> It is incomprehensible that God should exist, and it is incomprehensible that He should not exist; incompre-

hensible that the soul should be joined to the body, and incomprehensible that we should have no soul; incomprehensible that the world should be created, and incomprehensible that it should not be created.

The tension is tangible. There can be no secure and tranquil believing, nor any secure and tranquil non-believing either. Can the understanding then offer no solution here? The "incomprehensible," repeated six times, already indicates that it must not be sought in this direction. But Pascal has a high opinion of the understanding ("the order of the mind"):

Man is obviously made to think. It is his whole dignity and his whole merit; and his whole duty is to think as he ought. Now the order of thought is to begin with the self, and with its Author and its end.

Thought constitutes the greatness of human beings. But thought comes up against its own limits:

The last proceeding of reason is to recognize that there is an infinity of things which are beyond it. It is but feeble if it does not see so far as to know this.

Understanding that oversteps its limits is arrogant, but understanding which does not go to its limits is weak. Thought is part of human nature, but for that very reason it is limited. That also applies to science. Pascal does not deny the significance of science (that would have been difficult for a man who had been so intensively occupied with science all his life), but its claim to absolute validity.

How far can one get with thought? According to Pascal:

The greatness of man is great in that he knows himself to be miserable. A tree does not know itself to be miserable. It is then being miserable to know oneself to be miserable; but it is also being great to know that one is miserable.

Here Pascal comes close to a modern philosopher like Leszek Kolakowski, who says: "The philosopher serves to demonstrate that all empirical attempts to present our world as a rational meaningful order are doomed to failure. Existence does not have a meaning which we could infer as it were by experiment. Anyone who says that our existence has meaning moves outside the sphere of philosophical thought. That is certainly not forbidden, provided that you sense that you are leaving the sphere of reason and entering that of myth." To our ears the word myth may have the negative connotation of "fable," but not in Kolakowski's definition: "Myths are all spiritual constructions which give purpose and meaning to our life and our intellectual efforts. *Without them we could not survive.* It is essentially a religious matter."

So too Pascal. His remarks about "the miserable state" of human beings are not meant to establish this and leave it at that, but to make it clear that our life cannot rest on itself. He then says that it is no more than *reasonable* that people should make an effort to find a way which raises them above their hopeless existence:

> There are two kinds of people whom one can call reasonable: those who serve God with all their heart because they know Him, and those who seek Him with all their heart because they do not know Him.

In this quotation the key words occur twice: *their heart.* The understanding must arrive at the insight of where its limits lie, but that does not doom it to the irrational, the unknowable, but to the sphere ("the order") of divine love. Although Pascal says that it is reasonable to seek God, that does not mean that reason is the searchlight by which one can find him. The sphere of divine love lies outside the limits of understanding. Human beings are not excluded from this; on the contrary, it is where they find their true destiny. So whether or not they do so exclusively with their understanding, they must be able to react to this divine love. Pascal calls the organ with which they do this "the heart": It does not coincide with their understanding but does not conflict with it.

We should not spend time here over defining precisely what "the heart" means in Pascal. It is more than feeling. It is the core of human personality, of our attitude to life, the source of our actions. It is the center from which inclination springs. We should let Pascal speak once more in a fragment in which, without any trace of uncertainty or doubt, he describes where the way of the heart has brought him.

> The God of Abraham, the God of Isaac, the God of Jacob, the God of Christians, is a God of love and comfort, a God who fills the soul and heart of those whom He possesses, a God who makes them conscious of their inward wretchedness, and His infinite mercy, who unites Himself to their inmost soul, fills it with humility and joy, with confidence and love, who renders it incapable of serving any other end than Himself.

The words of this fragment suggest those of a unique document which was found by chance after his death, sewn into the lining of his jacket. It is known as the "Memorial." It is dated Monday 23rd November of the "year of grace" 1654, from half past ten in the evening to half past twelve at night. It describes in ecstatic words a deep mystical experience which came over him at that time. It begins with the words:

> Fire, God of Abraham, God of Isaac, God of Jacob, not of philosophers and scholars.
> Certainty. Certainty, Feeling, Joy, Peace.
> God of Jesus Christ.

No one knew of the existence of this document; he never discussed the experience with anyone, even with his sister Jacqueline. One feels almost an intruder when one reads it. It was never meant to be seen by anyone else. But I am glad that it has been preserved. It stresses once again what becomes clear in the *Pensées*: that this great scholar personally knew, encountered and served the God who is not a conception of philosophers and scholars but has made himself known, personally known, in the history of salvation as experienced by Jews and Christians.

ALBERT EINSTEIN

"Science without religion is lame, religion without science is blind"

Einstein (1879 - 1955) is without doubt the greatest scientist of our century. In this section I shall look briefly at his attitude to religion. The main source is again the biography by Abraham Pais which I mentioned earlier.

Einstein grew up in a liberal Jewish family, in which there was little talk about religious matters and principles. But he was instructed in Judaism by a distant relative and later by a history teacher at school. As a result, he went through a brief but intense religious phase at the age of eleven. He observed Jewish religious precepts strictly and, as he related later, composed hymns in honor of God which he sang enthusiastically on the way to school. Later he wrote: "It is clear to me that this lost religious paradise of youth was a first attempt to liberate myself from the 'only personal.'"

In some of his famous sayings, he makes mention of God. We have already come across "God does not play dice." Another is: "Subtle is the Lord, but malicious He is not." We must not conclude from this that Einstein believed in a personal God who can be subtle or throw dice. As we shall see, he did not. In these quotations we could just as well substitute "nature" for "God" without changing the meaning. Einstein uses the word as a somewhat ironical manner of speech (the Old One!), which perhaps also betrays how deeply he was under the impression of the majesty of nature. "If he had a God," says Pais, "it was the God of Spinoza." That seems to me well put, as will emerge in due course.

The main work by Einstein which gives us insight into his view of religion is a short article in *Nature* (1940), entitled "Science and Religion"; I shall try to convey its content. He finds it difficult, he says, to give a definition of religion, and so he prefers to ask:

> What characterizes the aspirations of a person who gives me the impression of being religious? A person who is religiously enlightened appears to me one who has, to the best of his ability, liberated himself from the fetters of selfish desires, and is preoccupied with thoughts, feelings, and aspirations to which he clings because of their *super-personal* value [...]
> What is important is the force of this super-personal content and the depth of the conviction concerning its overpowering meaningfulness, regardless of whether any attempt is made to unite this content with a Divine Being, for otherwise it would not be possible to count Buddha and Spinoza as religious personalities. If one conceives of religion according to this definition, a conflict between religion and science is impossible. For science can only ascertain what is, but not what should be, and outside its domain value judgments of all kinds remain necessary. Religion, on the other hand, deals only with evaluations of human thought and action; it cannot justifiably speak of facts and relationships between facts. According to this interpretation, the well-known conflicts between religion and science in the past must be all ascribed to a misapprehension of the situation.
> Now, even though the realms of religion and science are in themselves clearly marked off from each other, nevertheless there exist between the two strong reciprocal relationships and dependency. Though religion may be that which determines the goal, it has, nevertheless, learned from science, in the broadest sense, what means will contribute to the attainment of the goals it has set up. But science can only be created by those who are thoroughly imbued with the aspiration towards

truth and understanding, and the source of this feeling springs from the sphere of religion.

Einstein summarizes the reciprocal relationship between science and religion with the famous words written as the heading to this section.

But then he comes to what for him is the great difference between his view of religion and that of the historical religions, namely the concept of "God," who in the religions is seen as a personal God. Here for Einstein is the stumbling block which cannot be overcome, the most important source of present-day conflicts between science and religion. What for Newton formed the absolutely essential nucleus of his belief is for Einstein the stumbling block which must be rejected. He vigorously calls on the "teachers of religion" to have the courage to drop the doctrine of a personal God and advises them to devote themselves to the powers which can cultivate the good, the true, and the beautiful in humankind itself.

Why was Einstein so fiercely opposed to the idea of a personal God? The goal of science is to establish general rules, laws of nature, which have universal validity. That this is possible is a belief which cannot be proved, but has been increasingly confirmed, the more deeply human beings have penetrated into the natural order.

> The more a man is imbued with the ordered regularity of all events, the firmer becomes his conviction that there is no room left by the side of this ordered regularity for causes of a different nature.

He understands these "causes of a different nature" to be the human will and the divine will. What this amounts to is that he unhesitatingly holds fast to faith in a causal-deterministic view of the world. Given the discussion in the previous chapters, that should not be surprising.

Einstein ends with a passage which is so characteristic of his nobility of mind that I shall quote it in its entirety.

If it is one of the goals of religion to liberate mankind so far as possible from the bondage of egocentric cravings, desires, and fears, scientific reasoning can aid religion in yet another sense. Although it is true that it is the goal of science to discover rules which permit the association and foretelling of facts, this is not its only aim. It also seeks to reduce the connections discovered to the smallest possible number of mutually independent conceptual elements. It is in this striving after the rational unification of the manifold that it encounters its greatest successes, even though it is precisely this attempt which causes it to run the greatest risk of falling a prey to illusions. But whoever has undergone the intense experience of successful advances made in this domain is moved by profound reverence for the rationality made manifest in existence. By way of the understanding, he achieves a far-reaching emancipation from the shackles of personal hopes and desires, and thereby attains that humble attitude of mind towards the grandeur of reason incarnate in existence which, in its profoundest depths, is inaccessible to man. This attitude, however, appears to me to be religious in the highest sense of the word.

It is not difficult to recognize in this fragment the man of whom Abraham Pais says: "He was the freest man I have known. By that I mean that, more than anyone else I have encountered, he was the master of his own destiny. If he had a God, it was the God of Spinoza." And: "He once commented that he had sold himself body and soul to science, being in flight from the 'I' and the 'we' to the 'it.' Yet he did not seek the distance between himself and the other people. The detachment lay within and enabled him to walk through life immersed in thought. What is so uncommon about this man is that he was neither out of touch with the world or aloof." And finally, "According to his own definition, Einstein of course was a deeply religious man."

STEPHEN HAWKING

"For then we would know the mind of God"

Anyone who has come so far as to be appointed professor in Cambridge, England, at a very early age to the same chair as that once occupied by Isaac Newton; anyone who at the same time has made fundamental contributions in his professional sphere, cosmology, has published profound articles, and travels round the world giving lectures about it; anyone who in addition still finds time to give series of television interviews and to write a book in which he sets out in a winning way to explain the physics of the universe to a wide public, deserves great admiration for his intelligence, enthusiasm, and productivity. If the same man, afflicted by a rare disease, has a completely paralyzed body, as a result of which he is confined to a wheelchair and can only be moved around by others; if this disease has also completely deprived him of the capacity to speak, so that he can only make himself understandable with the help of an ingenious apparatus which produces an intermittent sound, then words fail to describe the power of his spirit. This man exists. His name is Stephen Hawking.

We have already met him several times in previous chapters. On those occasions I saw myself compelled to make some critical comments, and I shall have to make some more. However, that does not do away with the admiration I have just expressed.

His book is *A Brief History of Time*, which has long been on the best seller lists in many countries. In it he describes how, according to the ideas of modern cosmology, the universe as we now perceive it has developed from the moment that the original Big Bang took place. I shall not go into these ideas further here; that will be the subject of the next chapter. What I am concerned with now is the fact that in his book Hawking often speaks about God (the word God is the last word in the book) and the way in which he speaks about God.

In short, the question is this: Is it still possible or necessary to find a place, a little place, for God as Creator in the theory which tries to describe the origin and development of the universe? Is this hypothesis still needed? Hawking is skilled in maintaining the tension: Sometimes he seems to be positive, and later negative again. So he describes a congress in 1981 organized by the Jesuits in the Vatican with the aim of getting up to date with the most modern developments in cosmology and discussing them with expert participants. During an audience given by the Pope to the members of the congress, Hawking thought:

> I was glad then that he [the Pope] did not know the subject of the talk I had just given on the conference - the possibility that space-time was finite but had no boundary - which means that it had no beginning, no moment of Creation. I had no desire to share the fate of Galileo.

The poor Pope in the role of a child who still believes in Santa Claus while adults know better!

In Hawking's view the universe, human beings included, is a great and extremely complicated cryptogram which was - perhaps - once put together by God. If we arrive at a correct solution, in the middle a word of seven letters appears which - possibly - will disclose to us the name of the maker and his being. The solution of the riddle is kept for the brightest boy in the class, the physicist. God - if he exists - looks on with excitement. Will the quest succeed? Or are the seven letters "bad luck"?

Hawking is very confident that things will turn out well - in his estimation within the next ten years. The last sentences of his book read:

> However, if we do discover a complete theory, it should in time be understandable in broad principle by everyone, not just a few scientists. Then we shall all, philosophers, scientists, and just ordinary people, be able to take part in the discussion of the question of why it is

that we and the universe exist. If we find the answer to that, it would be the ultimate triumph of human reason - for then we would know the mind of God.

The boundless arrogance! Throughout their long history, human beings have yet to say a meaningful word on the question "Why do we and the universe exist?" This discussion can only begin when the scientists have finished. Then the Great Formula will be explained to human beings, and even simple souls will be enlightened. They may then join in the discussion. A pity about Jesus and the Buddha and all the others who weren't able to take part in it. I shall return in the next chapter to the question what the scientist can expect in the quest for God. Here I just want to bring out another aspect of Hawking's account. He says of the audience with the Pope at the time of the Vatican congress mentioned above:

> He [the Pope] told us that it was all right to study the evolution of the universe after the big bang, but we should not inquire into the big bang itself, because that was the moment of Creation and therefore the work of God.

If this were an accurate account of what the Pope said, there would be reason for amazement. The congress was concerned with the relationship between the church and science, which after Galileo was torn to shreds but is now beginning to improve somewhat. Here the Pope seems again to be making the same mistake as his predecessors by telling scientists how far they may go. But in reality, what the Pope said was this:

> Any scientific hypothesis about the origin of the world, like that of a primal atom from which the whole of the physical universe came forth, leaves open the problem of the beginning of the universe. Science cannot solve such a question by itself: This human knowledge must raise itself above science and astrophysics and what is

called metaphysics; the knowledge must come above all from the revelation of God.

So the Pope is not *prohibiting* the scientists from doing anything. He is simply quite confident that they cannot overstep a particular limit (but note the words "above all"!). Perhaps Hawking's "rendering" was influenced by his sense of a bond with Galileo, because they share the same birthday.

7 Physics: A Way to God?

Our little systems have their day;
They have their day and cease to be:
They are but broken lights of Thee,
And Thou, O Lord, art more than they
Alfred Tennyson

God and the New Physics is the title of a book by Paul Davies, the author of *Superforce*, which we have already met. The two books cover largely the same developments in modern physics and cosmology; the former goes a step further and connects these developments with God. Is there still room for him? Can he perhaps be found through science?

The author makes the purpose of the book clear in the Preface, in the sentence, "It may seem bizarre, but in my opinion science offers a surer path to God than religion." The two, science and religion, are set over against each other as two rival ways to God, the first of which Davies thinks superior. There is nothing against that as long as the "rival," religion, is taken seriously. In my view, there is no question of that in this book. Let me illustrate this with a couple of examples.

If the Church is largely ignored today, it is not because science has finally won its age-old battle with religion, but because it has so radically reoriented our society that the biblical perspective of the world now seems largely irrelevant. As one television cynic recently remarked, few of our neighbours possess an ox or an ass for us to covet.

I find such a person not so much cynical as obtuse. The commandment referred to is "You shall not covet your neighbor's house, or his wife, or his ox or his ass or anything that is his." It does not take the least intellectual effort to modernize ox and ass into a Rolls-Royce and a luxury yacht. You then immediately have a rule of life which anyone can read and understand, which people are vociferously encouraged to transgress, for example in advertising campaigns for cars. The desirability of the car being promoted is further stressed by draping it with the equally desirable "neighbor's wife." That such a rule is unpopular is really not because it contains "ox and ass," far less because science has overturned society to such an extent. Desire is the goal of commerce. Here is another passage:

> Scientists also are searching for a meaning; by finding out more about the way the universe is put together and how it works, about the nature of life and consciousness, they can supply the raw material from which religious beliefs may be fashioned. To argue whether the date of the Creation was 4004 BC or 10,000 BC is irrelevant if scientific measurements reveal a 4.5 billion year old Earth. No religion that bases its beliefs on demonstrably incorrect assumptions can expect to survive very long.

Here again we have a contrast between the scientist who provides the raw material for a belief and the stupid believers fighting over outdated problems. I'm also well aware that in a particular variant of Christianity, fundamentalism, these problems are still not thought to be outdated, but even for this variant it is ridiculous to claim that such questions are the *basis* on which faith rests.

Davies is also fond of quoting from a book by the physicist Hermann Bondi, who, as he himself comments, does not have a good word to say for religion. For example, he makes this kind of comment:

> Nothing, however, is further from the believer, any believer, than this elementary humility. Everyone in his power (who nowadays in a developed country tend to

be confined to his children) must have his faith rammed down their throats. In many cases children are indeed indoctrinated with the disgraceful thought that they belong to the one group with superior knowledge who alone have a private wire to the office of the Almighty, all others being less fortunate than they themselves.

Davies certainly tones down such statements (to some degree), for example by commenting, "Of course, not all religious people are fanatical zealots," but they are nevertheless there.

A proven technique of discussion is also to put words in the mouth of one's opponents, to attribute views to them in which they certainly would not recognize themselves, and then go on to challenge these views. The book is full of sentences like "Christians assert," or "A theologian would comment," while in my wider surroundings I do not know any Christians who would say this sort of thing or any theologians who would want that kind of remark to be attributed to them. In the chapter entitled "Miracles" the author spends pages on a discussion between a fictitious "believer" and a fictitious "skeptic" on the subject. From the believer's first sentence, "In my opinion, miracles are the best proof that God exists," he has one silliness after another, so that the "skeptic" has little difficulty in winning the contest at every point. The skeptic also makes use of very strong language; for example, when he describes the evangelists (Matthew, Mark, Luke, and John) as "a lot of superstitious zealots with a vested interest in promoting their own brand of religion."

Davies argues for his own belief as follows: "The tremendous power of scientific reasoning is demonstrated daily in the many marvels of modern technology. It seems reasonable then, to have some confidence in the scientist's world-view also." Miracles prove nothing about the existence of God, but miracles of another kind, like computers and atom bombs, must make it plausible that the brains behind them should also have a good view of God and human beings.

But enough of that. I now want briefly to give the argument of the book. Davies first formulates what he calls the Four Great Questions of existence:

1. How did the universe achieve its organization?
2. How did the things the universe consists of come into being?
3. Why does the universe consist of the things it does?
4. Why are the laws of nature what they are?

Note that these are the main questions *of existence*. If people know the answer to them, existence no longer has any secrets. The possibility that these could perhaps be the main questions of *physics* is never discussed.

The present state of affairs is that the first question has been fairly satisfactorily answered. At least the fact that the state of things is now better ordered and organized than it was at the beginning of the Big Bang, when complete chaos reigned, is no longer in conflict with the laws of nature, as was long thought. Some progress has been made in answering the second question, while the solutions to the third and fourth questions are at a purely speculative stage.

Let us now follow briefly the history of question 2. It is clear that bodies like stars and planets are formed from the primal gases (predominantly hydrogen and helium), which in turn came into being shortly after the Big Bang. Here Davies says that the Bible is not clear about the question whether the material of which the stars, the planets, and our own bodies are made already existed before the event of creation or not. He thinks that believers have no other choice than to suppose that God also created this material, because otherwise he would not be omnipotent (this is the first of many occasions on which the author uses the term omnipotence, and I shall return to that). At any rate, had God not created matter, he would have been limited in his work because he had to work with material which was already there.

Secondly, it has been demonstrated that matter as we know it need not have been created, but could have come into being in a "natural" way, from energy. From an energy quantum two particles can come into being which are each other's "anti-particles," like an electron and a positron (an electron with a positive charge). If the matter known to us came into being in this way, then until recently it was a great mystery why the universe as we know it consists almost entirely of matter (protons, electrons, etc.), and where the antimatter (antiprotons, positrons, etc.) that goes with it is. Some light has recently been shed on the puzzle of this imbalance: At the very high temperatures at which the process of materialization takes place, there is a very slight imbalance. There is about one billionth more matter than antimatter. Directly after being formed, the particles of matter which came into being from the radiation united with the relevant particles of antimatter to form energy, which we can still observe as the heat of the radiation which comes to us from the universe. The temperature of this radiation fits the calculations. Any matter that is left over is that billionth part; the whole observable universe consists of it. Having got this far, Davies remarks:

> The question of the origin of matter illustrates a fundamental problem that faces any attempt to deduce the existence of God from physical phenomena. What once seemed miraculous - the appearance of matter without antimatter - perhaps requiring a supernatural input of the big bang, now seems explicable on ordinary physical grounds, in the light of improved scientific understanding.

The argument so far is typical of what follows. First, it is stated that "believers" or "theologians" or "Christian doctrine" are compelled to think that God created matter directly, from nothing. Then Davies demonstrates that science does not need this supernatural intervention, because matter arose naturally from energy. And then he again introduces the "theologians" who, forced back on the following line of defence, now say:

The processes described here do not represent the creation of matter out of nothing, but the conversion of pre-existing energy into material form. We still have to account for where the energy came from in the first place. This surely requires a supernatural explanation.

But as you will already have guessed, that is not necessary at all: The total energy which the universe contains is calculated as being nil! And something that is nil does not need to be created. Energy arose from empty space. In that case we only need to explain *how* energy can arise from empty space. In the meantime, a reasonably satisfactory theory of this has been produced, based on the laws of quantum physics. Then the following question immediately arises: But where does this empty space, and time associated with it, come from? Could it not be that God at least created that? In Davies' words:

It could be argued that science has still not explained the existence of space (and time). Granted that the creation of matter, for so long considered the result of divine action, can now (perhaps) be understood in ordinary scientific terms, is it only by an appeal to God that one can explain why there is a universe at all - why space and time exist in the first place, that matter may emerge from them?

Before Davies discusses this question, which, as we shall see, is still far from being answered and about which for the moment one can only speculate, he makes a remarkable digression. At this point, he takes a great deal of trouble to disable the so-called "cosmological argument" for the existence of God which was developed by the medieval philosopher Thomas Aquinas and was later consolidated by others, including Leibniz. He then demonstrates that the concept of "cause" used in this argument conflicts with quantum physics and that the term "time" which is used does not correspond to what the theory of relativity teaches us about that. You might think that this is not really surprising. He does also say that this cosmological argument was already vigorously disputed by David Hume, Immanuel Kant, and Bertrand Russell (though these were hardly prominent spokesmen for Christianity), but not that already Pascal saw nothing in it. For believers, Pascal says, this kind of proof is unnecessary because they know God in other ways, and unbelievers are rightly not convinced by it. The fact that the name Pascal does not occur at all in Davies' book is, of course, significant.

But back to the origin of space-time. A theory of the "spontaneous" origin of space-time does not yet exist, but there are ideas about what such a theory should look like. The solution is sought in the application of quantum theory to gravity. In particular, this theory should make it possible for space-time to be generated and annihilated spontaneously and without a cause, just as ordinary quantum physics makes it possible for particles to be formed and annihilated spontaneously. We have not got that far yet, but if that does happen it will already be clear that the size of such a piece of space formed spontaneously would be extremely small: billions of billions of times smaller than an atom. It would then have to be explained how such a mini-universe can develop to the proportions we know. Theories have already been developed for this process of expansion (the "inflation scenario") which are already quite acceptable.

If all this works out satisfactorily we have the answer to question 2, "Where does the universe come from?" The answer is, "From nothing." This answer agrees only superficially with the Christian doctrine of *creatio ex nihilo* (creation from nothing) because this is not a creation, for which a creator would be necessary, but the *origin* and subsequent development of the universe which is to be described wholly by means of the laws of nature.

That leaves two questions. First, question 3. Why does the universe consist of the "things" it does (electrons, protons, etc.)? We know in general how an electron or proton behaves when it is by itself, but we have no clear idea why they are electrons or protons and not other particles. We are still far from arriving at such a theory, but here too Davies is full of hope. He believes that it is a matter of time. The solution to this question must come from the theory of supergravity, which I shall not discuss further here. Davies describes this vision of the future like this:

If supergravity is fully successful, it will tell us not only why there are the particles that exist, but also why they have the masses, charges and other properties that they do. All of this will follow from a magnificent mathematical theory that will encompass all of physics in one super-law.

Have we then finally settled things? The scenario of the "universe from nothing" claims that all that we need are the laws of nature (or, even more attractively, the one super-law) - after that the universe can get on by itself, including creating itself. Has God now definitively vanished from the scene? No, for there still remains the last question, which Davies calls "the one fundamental question of existence." Why are the laws of nature as they are? Did God perhaps think them up? Did God, so to speak, choose which super-law he would use in order to set the universe working? There is still one hope that we can also get round that. To begin with, the choice which God had cannot be infinitely great; the universe that he chooses has to be *logically consistent*. That is already a limitation to his power, since evidently he cannot choose everything. Things would become very much worse if it should emerge that the super-law that we discovered, and thus the universe as it is, is the only logical possibility. Then the omnipotent God would shrivel into someone who did not have to choose, and then he could finally be dismissed as creator.

So does this philosophy of a unique physical solution to the fundamental logical mathematical equation of the universe deny the existence of God? Indeed not. It makes redundant the idea of God-the-creator; but it does not rule out a universal mind existing as part of that unique physical universe: a natural, as opposed to supernatural God.

But the existence of such a God is already uncertain. As Davies explains elsewhere in the book, "mind" means "organization." Now the second law of thermodynamics specifies that disorder, disorganization, must constantly increase in the universe, so that the universe ultimately dies the "death of disorder." The degree of organization in the universe, and thus God, goes on decreasing until there is nothing left. God dies by his own second main law, which he could not revoke because he had no other choice at the beginning.

That is a summary of Davies' argument, which I have commented on as I have outlined it. I want to go more closely into five points.

1. God's "omnipotence"

Time and again in his argument, Davies constructs a contrast between God's omnipotence and one of his other "properties." If God is this or that, he cannot at the same time be omnipotent. Here Davies uses a kind of mathematical definition: Someone is omnipotent who can do *everything*. If one can show that God cannot do a thing, then he is not omnipotent. That seems to be a logical definition, but it isn't. It's absurd. I've known that since my earliest youth, when my "unbelieving" friends asked in triumphant merriment, "So God's omnipotent, is he? Then can he make a stone so big that he can't lift it?" Precisely the same paradox is known in mathematics under the name of Bertrand Russell's paradox. It runs: Let R be the set of all the sets which do not contain themselves as an element. Does R then belong to the set? For anyone who finds this too abstract Russell described the problem as follows: The village barber shaves all the men of the village who do not shave themselves. Does the barber shave himself? On reflection one can see that both the answers "yes" and "no" are wrong. The question, like that of the omnipotent stonemaker, leads to an absurdity. The reason is the same in both cases: Evidently one cannot use the concept *everything* sensibly in this way.

The use of the term "omnipotent" in this way is thus in itself meaningless and moreover has nothing to do with the way in which the Bible talks of God. Here is another example by way of illustration. In a chapter on "time," Davies constructs a new paradox when he is discussing the relationship of God to time. Briefly, it amounts to this. There is an alternative: Either God exists in time or he stands outside it, is timeless. In the first instance, he is subject to the scientific laws for time, and (here we have it again) in that case he cannot be omnipotent. In the second instance, he cannot be a personal God, since everything that a person does, like reflecting, feeling, making plans, take place in time. Hence the conclusion: God cannot be omnipotent and a personal God. But it was hardly necessary to invent the theory of relativity to arrive at this conclusion. It can be found literally in the Bible, pointed out by the apostle Paul, who two thousand years ago wrote, "Though we are unfaithful, he remains faithful, since he cannot deny himself." There are some things that God cannot do because he is a person who is completely faithful. To put it even more strongly: This "omnipotent" can sometimes seem to be *less* than his creatures, who usually find no difficulty in being unfaithful. Davies' whole use of the concept of "omnipotence" rests on the quicksand of an absurd definition which has nothing to do with the content of the biblical word (which of course is a faltering translation of the almost untranslatable Hebrew word *Shaddai*). Anyone who wants to know what that word can mean should consult an elementary book on theology.

2. The God of the gaps

The "theologian" who is introduced into Davies' argument plays the following role there. At every stage at which scientists have taken another step forward in explaining something previously unexplained, he says, "All right, granted, so far (unfortunately) I have to go along with you. But you won't find a solution to the following problem; there you will certainly need God as an explanation." He is constantly on the defensive, is constantly thrown back on the following line of defence. He uses God to fill in the "gap" of everything that has yet to be explained. The gap gets steadily smaller. He continues ever more desperately to look for a place for God somewhere. When the gap is at last definitively closed, he surrenders. Then God is no longer needed.

In this connection, Davies writes a sentence with which I am in full agreement, namely, "If God is to be found, it must surely be through what we discover about the world, not what we fail to discover." I agree in a sense which was probably not intended by Davies, namely if we understand "the world" in the widest possible sense, as the world of all human experiences, of which physical perceptions and theories are only a part. I would go a step further: It is even extremely improbable that we should encounter God in the latter experiences, *by definition*: It is in the nature of science that this cannot be. Let me try to explain.

The fact that anything like science is possible at all is because human beings who look at the reality around them soon see that chaos and arbitrariness do not prevail there, but order and regularity. The sun rises every day, and ebb and flow have their fixed times by which one can set a clock. A stone which falls downwards one day does not rise into the sky the next; a car which one puts in first gear will go forwards today and also tomorrow. We are so used to this that we find it completely obvious, but it isn't. As far as I'm concerned, it can confidently be called a miracle. The physical reality around us is dependable and that is a basic condition of our existence. Someone put it like this:

Reality is dependable. It is not a haunted house or a bizarre fairy tale. We can depend on it. We can orient ourselves in it, feel secure in it, and make plans for its and our future. Its habitability depends on its knowability, and this knowability is that of a universe governed by law.

What physicists do is to penetrate this regularity. They describe physical reality in laws which are as simple as possible and which cover as large groups of phenomena as possible. To a large extent that works: to the deepest depths of the atom, to the farthest corners of the universe, to the beginning of the Big Bang. As long as it works, there is no reason for anyone to bring God in: certainly not unbelieving physicists, but not believing physicists or theologians either. For all of them, there is only reason for amazement and wonder that the order which we can already perceive superficially with the naked eye is anchored so deeply in the cosmos. (The Greek word *kosmos*, of course, means order.) There is no reason at all on any step forward to exclaim triumphantly that there, too, we do not need God again.

What about God, then? The quotation above, about the trustworthiness of reality, is not - as you might think - from a physicist, but from a theologian. It comes from Herman Berkhof's *Christian Faith*, from the chapter on creation. I have changed just one sentence, the first. This in fact reads (in somewhat abbreviated form):

> Createdness by God who is faithful means that reality is dependable.

That is an example of how real theologians (and not Davies' talking puppets) talk about creation. There is no suggestion of rivalry between God and the laws of nature; that God should be driven from the steadily growing area in space and time which according to our perception is dominated by the laws of nature. The laws of nature reflect God's dependability, which as a result has made possible a livable existence for human beings.

Of course, this is a *statement of faith*, which is not founded on the laws of nature themselves. We do not as it were see God's dependability there, but only get a glimpse of it if we have learned to know it elsewhere in the world of human experience (an important part of the rest of this book is devoted to the question whether and how that is possible). That is precisely the way in which Pascal formulated this three centuries ago. Only in this way is God "necessary" as creator. As a filler of the gaps in physicists' understanding about the Big Bang (so far!) God can be dispensed with.

3. Science and belief in creation

If you look at the laws of nature in the light of biblical belief in creation in the way that I have just described, has any tension between the two definitively been removed? If we are to be able to answer this question, we must first go rather more closely into the content and significance of the creation stories which are related in the first chapters of the book of Genesis. I deliberately use the plural "stories," because one of the first things that strikes one on reading the first two chapters of Genesis is that the creation of human beings in particular is described in two different ways. If one took them as a literal account of an actual event in the past, the two stories would directly contradict one another. In that case there are only two possibilities. The first is that the writer was so utterly stupid that he did not notice that he was already contradicting himself on the first two pages of his book. The second - and nowadays this is the view that is accepted almost universally - is that the stories do not in any way claim to be an eye-witness account of someone who was there, or a dictation from God, who was there, to a writer who wrote it down obediently, but that these are *stories*. Note that I say stories and not fables. These are stories which have a meaning, which try to make something clear about the relationship between human beings and their world, between human beings and God, their creator.

The stories were written down some centuries before the beginning of our era, during a nadir in the existence of the Jewish people, the Babylonian exile. By then the people already had a long history behind them in which God, from the patriarch Abraham onwards, had become involved with them, had pointed out the way, had shown himself to be the Liberator, had given them the Torah, and had turned to them through his prophets when they tended to leave the way. The creation stories were conceived on the basis of these experiences with God in history as "extrapolations back to the beginning" in which, in figurative language, expressed in the world-view of the time, the purpose of God with his creation and with human beings, the crown of creation, was expressed. For anyone who can read them - and theologians are important helpers in teaching us to read them they have a rich content. I have already indicated just one aspect of this content above (the dependability of physical reality as a reflection of the dependability of God). There is no room here for a detailed account: for that I would refer the reader to the book by Herman Berkhof, *Christian Faith*, which I have already mentioned, or to the very readable book by William Van der Zee, *Ape or Adam?*.

It will be clear that the creation stories, understood in this way, cannot be the source of scientific information. They tell us nothing about electrons and quarks, about the origin of the milky way or events during the first second. They give no answer to questions like Davies' question whether God created matter directly, or energy, or space-time or the natural laws. It is reserved for the physicists to occupy themselves there in complete freedom and with deep amazement. Their results cannot threaten belief in creation. On the contrary. If "a magnificent mathematical theory should ever encompass all of physics in one super-law," believing physicists will be just as delighted as their non-believing colleagues, because in it they will find the sublime simplicity of God which is also expressed in the "super-law" of God for human existence: "God is love. Let us love one another."

Are there, then, no more problems? Perhaps yet one more looms right at the end, when the whole scenario that Davies has sketched out comes to the end that he hopes for and believes in. Let us remind ourselves: The penultimate step in it was that there is one natural law on the basis of which the universe can develop itself from nothing. The last step is that this natural law is the one (logical) possibility. God had no choice. That indeed seems in a way threatening to creation faith. For although we saw that the biblical creation stories do not seek to give any information about the question what was actually involved in the act of creation in a material sense, that does not alter the fact that there must be room for it somewhere.

For the moment, I cannot lose much sleep over that. We have seen that a considerable part of the scenario is still purely speculative. At every step, Davies reports how things "must" work out. By "must" he means that only a physical explanation is possible in each case. The theory of quantum gravity "must" make it possible for space-time to have been generated spontaneously. And so on. If the super-law is found, that "must" be the only possibility. That must be the case because, as he declares somewhere, "the scientist gives priority to a world which works in accordance with the laws of nature." That is the presupposition from which he begins, which I confidently dare to call a faith.

As can be inferred from what I have already said, I would be quite content if the whole scenario could be worked out up to and including the penultimate step. I would have some difficulty with the last step, as now envisaged. But even that is less serious than it seems, because this clashes only with God's omnipotence in the sense described above, and that need not be thought too serious. And in all this it is still worth keeping in mind the attractive lines of Tennyson which I put at the head of this chapter, and which Berkhof used as a motto for his book on theology (!). "Our little systems have their day." Our little thought systems, like theology - and, of course, physics.

> They have their day and cease to be:
> They are but broken lights of Thee,
> And Thou, O Lord, art more than they.

4. Overstepping bounds

We should now pay some attention to Davies' remark that "scientists provide raw material for a belief" and his suggestion that "science forms a more certain way to God than religion." Given that his scenario provides a very detailed description of how science leads us to the final insight and roughly what that insight should look like, the final result can already be judged now with reasonable certainty. It looks like this. The universe created itself from nothing. The super-law can explain everything from the beginning. This super-law is the only one possible. If God then still represents anything, at most he can be present as a mind in that unique universe, and along with the universe he will ultimately die the death of disorder.

I cannot help finding this final stage of this "scientific way to God" extraordinarily poverty-stricken, to put it too mildly. It doesn't mean anything to me at all. If this is meant to be the answer to "fundamental questions of existence," I cannot make anything of it. There isn't even the beginning of an answer to the simple question "How must I live?" (and is that not perhaps the fundamental question of existence?). On the day that the Great Formula sees the light of day, it will change little in the human world, just as the first landing on the moon or the theory of electro-weak force did not produce any perceptible improvement in human behavior. If the way to God which is shown by religion represents even less than this, then religion indeed has *nothing* at all to offer.

It must be stressed here that views like those represented by such authors as Paul Davies and Stephen Hawking are quite extreme in the contemporary world of physics. Earlier on we met physicists like David Bohm and Max Jammer who let us hear quite a different voice. A man like Abraham Pais, one of the few people to see the full breadth and depth of twentieth-century physics, answers the question whether he thinks that physics is almost over: "I have no sympathy at all with this view. No sympathy at all. There are so many unsolved problems at a basic level that I find it quite premature to say that we've almost finished. Personally I don't believe that at all."

Then why have I paid so much attention to the opinions of Davies and Hawking in particular? Precisely because these two have expressed their views in popular books which are sold all over the world in editions of millions. Evidently they meet a great need among a non-scientific public. Davies himself also notes this:

> Many ordinary people too, searching for a deeper meaning behind their lives, find their beliefs about the world very much in tune with the new physics. In giving lectures and talks on modern physics I have discerned a growing feeling that fundamental physics is pointing the way to a new appreciation of man and his place in the universe.

In my view these writers, who meet these needs in the way which Davies describes, are guilty of overstepping bounds: Physics is made a pseudo-religion of which the physicists are the priests. Believers are then given this kind of prose:

> If Euclidean space-time stretches back to infinite imaginary time, or else starts at a singularity in imaginary time, we have the same problem as in the classical theory of specifying the initial state of the universe: God may know how the universe began, but we cannot give any particular reason for thinking it began one way rather than another.
>
> On the other hand, the quantum theory of gravity has opened up a new possibility, in which there would be no boundary to space-time and so there would be no need to specify the behaviour at the boundary. There would be no singularities at which the laws of science broke down and no edge of space-time at which one would have to appeal to God or some new law to set the boundary conditions for space-time. One could say: "The boundary condition of the universe is that it has no boundary." The universe would be completely self-contained and not affected by anything outside itself. It would neither be created nor destroyed. It would just BE. [Hawking]

That is language with which most physicists have the utmost difficulty. For outsiders it is language compared with which old church Latin was still crystal clear. It comes closest to magical incantations, the language of the magician in the most primitive forms of religion. I think that people do not so much "recognize their own ideas" here but that it is precisely what they are looking for: magical language which can put them in a kind of mystical trance. Of course the magicians legitimate themselves with *miracles*:

> The tremendous power of scientific reasoning is demonstrated daily in the many marvels of modern technology. It seems reasonable then, to have some confidence in the scientist's world-view also. [Davies]

It is only fair to note here, although the matter is so overfamiliar and over-emphasized that I hardly dare to do so, that the opposite overstepping of bounds, in which religion makes the claim to be a (pseudo-)science, is just as reprehensible. The condemnation of Galileo's ideas by the church is the most famous example. Even today, fundamentalist Christians think that the age of the earth is to be derived from the information in the Bible. Fundamentalist Islam is another terrifying image of how things should not be. The marking out of ground between faith and science will remain a problem, but the tensions at the frontier could be reduced by reasonable discussions between reasonable people, above all if they have "our little systems have their day" on their banners.

5. Is the believing physicist a schizophrenic?

In the previous section, I came to the conclusion that the results of modern science are not in conflict with, for example, biblical belief in creation. Anyone who takes both seriously is far from being a schizophrenic. I also argued that for anyone who wants to find God, science is not the first way, and indeed is probably an impossible way. The question of God, who he is and what he does, must be answered elsewhere in the world of human experience. In that sphere the physicist is no different from the baker or the bus driver, and there should be no reason for us to investigate the question how the physicist in particular proceeds in this area. I could end the book here.

But there is a reason why I have to go further. There is another reason for the charge of schizophrenia which we heard earlier from Simon Van der Meer and in which, as we shall see, he is not alone. It is that in the sphere of physics and in that of religion, two attitudes are asked of the physicist which are so completely different, indeed which are in such complete conflict with each other, and therefore so irreconcilable, that they cannot be combined in one person. If that happens, such a person becomes schizophrenic: If he is Dr. Jekyll, Mr. Hyde must be switched off, and vice versa. Davies puts it like this:

> The scientist and the theologian approach the deep questions of existence from utterly different starting points. Science is based on careful observation and experiment enabling theories to be constructed which connect different experiences. Regularities in the workings of nature are sought which hopefully reveal the fundamental laws that govern the behaviour of matter and forces. Central to this approach is the willingness of the scientist to abandon a theory if evidence is produced against it.
>
> In contrast, religion is founded on revelation and received wisdom. Religious dogma that claims to contain an unalterable Truth can hardly be modified to fit changing ideas. The true believer must stand by his faith whatever the apparent evidence against it. This "Truth" is said to be communicated directly to the believer, rather than through the filtering and refining process of collective investigation. The trouble about revealed "Truth" is that it is liable to be wrong, and even if it is right, other people require a good reason to share the recipients' belief.

The contrast is clear. When physicists are occupied with physics, they are critical, look for proofs, and believe something only if it is indubitably proved. Then in their leisure time, for example on Sundays, they turn a switch. They set their understanding to zero and their gaze on infinity. They exclude any critical sense. They need no proofs: They open their mouths and swallow; truths and absurd dogmas disappear down their throats, just as when as children they had to swallow a spoonful of cod-liver oil at bedtime. Their mothers used to stand there and say, "It's fishy, but it's good for you. You'll get a peppermint to take away the fishy taste." So too with God. If you swallow his dogmas indiscriminately, plus the revealed Truths which "are communicated direct to believers," (which so to speak rain down from heaven) and if you ask no questions, then you will be well rewarded: a good feeling while you live, and heaven when you die.

As usual, Davies' view of things is a complete caricature. Because he is not the only one to take this view, I feel it necessary to go into it more closely, as I shall do in the next few chapters.

8 How Do We Prove Anything?

"Now faith is the assurance of things hoped for, the conviction of things not seen" *Hebrews 11:1*

A remark which can often be heard, especially in scientific circles where observation and proof are highly thought of, is: "I believe only what I have seen with my own eyes or something of which clear proof can be given."

Two things can be said straight away to challenge this statement. Since I saw an attractive and confusing exhibition of optical illusions, I trust my eyes less than I used to. Moreover, Blaise Pascal has already pointed out that there are things in which everyone believes without there being a trace of evidence for them, like, "Tomorrow is another day," and, "One day we all die!" We believe that because - so far as we know - it has always been the case and will continue to be so. And finally, what is clear proof anyway?

In mathematics proofs are produced in the strict, exact sense of the word. Once the proof has been given, no discussion or doubt is possible any longer. The sum of the three angles of a triangle is 180 degrees. That has been proved once and for all; it's settled. But mathematics has to begin somewhere, to choose starting points. These starting points are called axioms. *They are unprovable.* The result of the three angles of a triangle is connected with the axiom: The shortest distance between two points is a straight line. If we replace this axiom by another, then the mathematics changes. There are different kinds of mathematics, like Euclidian and Riemannian geometry. Mathematics is unique in its method of demonstration. In other areas of human experience - including science - things are different.

Suppose we look at jurisprudence, where the concept of proof plays a major role. The guilt of the accused has to be proved. How is that done? The proof is the task of a public prosecutor who produces it by presenting *evidence*. The evidence is almost always indirect, because the crime itself is very rarely witnessed by one or more witnesses. It involves fingerprints, footsteps, bullets which have come from a particular gun, alibis, motives (who has an interest?), and suchlike. It involves the trustworthiness of the witnesses for the prosecution and the defence and how convincing they are, and it is the role of the prosecutor and the defence lawyer to test them critically. The prosecutor constructs a "case" from the total evidence: a model in which the pieces of evidence fall into place and make up a single picture which "proves" the guilt of the accused. Whether that proof is given or not depends on the impression that the total evidence makes on those who have to pass judgment: the jury or, in some countries, the bench of judges. They weigh the evidence and arrive at the verdict "guilty" when the case is established "beyond reasonable doubt." What is thought reasonable and what not is a subjective matter: Some people are more quickly convinced than others. Absolute, objective certainty cannot be attained, and we may take as a warning those numerous detective stories which demonstrate how an apparently watertight case can rest on quicksand. Here, too, it often happens that a case has to be reopened, sometimes years later, because a previously unknown piece of evidence has broken apart the whole chain of proof.

In daily life, we make up our minds about what we think to be true and what not, in roughly the same way as in a law court. Thus, for example, almost all of us believe that on the other side of the Atlantic there is a land called Europe; we have a definite picture of it, although most of us have never been there. How did we arrive at that belief? The answer must be that all the evidence, produced by a series of eye-witness statements, carries sufficient power of conviction for us. The first witness who told us about Europe was the teacher in elementary school who pointed out Europe on a globe and told us remarkable things about it. The teacher had never been there either, but we believed him or her unconditionally; this indicates that the personality of the witness and the authority that he or she radiates are of great importance for the persuasiveness of their statements. In our later life, belief in the existence of Europe has been steadily strengthened by the production of constantly new evidence. We read accounts of the Second World War, saw acquaintances or members of our family move to Europe and received letters from them. And last but not least, every travel bureau tells us that for the sum of around $ 500 - by no means a small amount but within the reach of many people - we can buy a return ticket to Amsterdam. The total evidence is more than enough to establish the existence of Europe beyond any reasonable doubt for almost everyone. *And if we so wish, if we devote enough money and trouble to it, we can confirm that belief by our own experience ("make it true").*

The basis of science is the evidence, which is provided by the observations of researchers. These report the results of their experiments in the form of publications: *They bear witness to their experiences.* This witness is - in principle - not accepted at face value by the scientific community. The requirement is that the experiment should be capable of being reproduced, being repeated. In other words, the experiment must be described so accurately that other scientists *who have the necessary means at their disposal* can repeat the same experiment so as to produce the same result.

A good example of this is the recent controversy over "cold fusion." This is a bit of modern physics which, because of its possibly far-reaching consequences for providing energy in the world, has made the front pages of all the newspapers. Two researchers from the University of Utah reported to a scientific journal, and simultaneously through a press conference to the world press, that they had succeeded in bringing about the fusion of two hydrogen nuclei at room temperature by a relatively simple experiment. They presented the special features needed to make their story acceptable: A marked development of heat was noted, and in the process neutrons and gamma rays were released, which seemed to have the energy required by the theory. The consternation provoked by this communication was caused above all by the fact that up to that moment scientists thought that this process could only take place at temperatures of many million degrees ("hot nuclear fusion") and that for decades attempts had been made to bring this process under control with extremely expensive projects. No wonder that immediately afterwards attempts were made to repeat Pons and Fleischmann's experiments in numerous laboratories. The results so far have been far from conclusive, and very confusing. As long as this goes on, the experiment is unacceptable as evidence to scientists.

For the experiment described here, the requirement that it should be capable of being reproduced is relatively easy to meet because the experiment is relatively simple. Things are different with most results in modern physics. They are usually obtained with the aid of apparatus which is so complicated and so costly, and by researchers with such specialized skills, that it is beyond virtually anyone's reach to verify the results. A good example of this is the work of Carlo Rubbia, Simon Van der Meer, and their colleagues, which I mentioned earlier, by means of which the existence of the W and Z particles were demonstrated with the help of the gigantic apparatus of CERN in Geneva. This experiment has now been repeated, by researchers at Stanford University, but for the moment that will be that, because no one else has the necessary facilities. The possibility of reproducing experiments like these has now virtually become an illusion.

It might also be added that the proof which is adduced for the existence of the W and Z particles would be of dubious value to a court's jury. For example, the (devil's) advocate could remark that if the evidence proved positive it would almost certainly lead to a Nobel prize for the researchers. Could researchers with such great interests at stake be unprejudiced? Do we have to take them at their word? Within the scientific world they were indeed taken at their word, and the expected Nobel prize was granted almost immediately. So why were the results of the experiment accepted universally and without criticism in this case, whereas from the start the findings of Pons and Fleischmann on cold nuclear fusion came up against worldwide skepticism and unbelief? One can only guess, but at least two factors seem to have played a role here. In the first place, the results of the CERN experiment agreed with the predictions of the theory and were therefore expected; the "cold nuclear fusion" results conflicted with the existing theory and were therefore unexpected. In the second place, the CERN group consisted of a large number of researchers of world-wide repute and eminent skill in their subject; with the two cold fusion researchers that was the case to far less an extent. In science, the important thing is not only what is reported and how, but also who does it: Authority plays a role.

So far I have been discussing what I called the basis of physics: the experimental evidence that is presented by the researchers. A second thing is inextricably part of it. As though before a jury, the evidence is collected with the aim of constructing a "case," i.e. a theory, a model of a part of physical reality, in which all the pieces of evidence fall into place. As long as the pieces fit, the theory is thought to be correct. It describes the whole of the perceived phenomena in the sector to which they relate. From a good theory follow *predictions* which forecast the result of an experiment that has not yet been carried out. As long as things continue to go well, the theory holds. As soon as a new and convincing piece of evidence conflicts with the theory, the theory has to be revised. The trial has to be reopened. Usually that leads to relatively small adjustments, but sometimes a completely new conception is necessary. So at the end of the nineteenth century, Newton's laws of motion were thought to be correct until it was demonstrated that they were in conflict with particular experimental facts. From this the theory of relativity and quantum mechanics were born; they were completely new conceptions which included Newton's mechanics as a limit case, an approach which is good under certain conditions.

Scientific theories have a *temporary* character; they work as long as they are satisfactory, and are replaced by better ones as the need arises. I have already demonstrated in connection with the rise and fall of the deterministic world-view that as a result we have to be extremely careful about drawing philosophical conclusions from a particular scientific theory.

To sum up: Scientific evidence is adduced by researchers, who in so doing bear witness to their experiences. That witness can in principle be confirmed, because the experiment has to be capable of repetition. In practice, the possibility of confirming findings is very limited, and is restricted to experts who must have the necessary means. The acceptance or rejection of the evidence is a process that takes time and must develop into a consensus. Subjective factors play a part here.

The total evidence, accepted by a large majority of scientists, is described in theories, models which have a finite existence and which have to be constantly adjusted or replaced by new ones. The thought that an end might be put to this, indeed quite soon, by the formulation of a definitive, "all"-embracing, theory certainly finds no support in the history of science and can still be described as a belief without a proper foundation.

Now we come to the question, "Can the existence of God be proved?" In the past the church, especially in scholastic thought, tried to produce so-called proofs for the existence of God, abstract arguments which had to demonstrate the existence of God beyond reasonable doubt. Pascal already said - to sum him up briefly - that these were useless. Those who did not believe would not be convinced by them, and those who did had no need of them.

Here we shall approach the question from another angle and not understand the term "proof" as an abstract, logical reasoning with no room for maneuver, of the kind that is to be found in mathematics, but as the collecting of evidence to make something acceptable "beyond reasonable doubt." Just as the public prosecutor fits the statements of the witnesses into the construction of a proof for his case in a court of law, or as scientists collect their evidence to test a theory, so we can ask: What is the evidence which makes the existence of God acceptable or not? How convincing are the declarations of the witnesses?

First, the writers of various books of the Bible lay great stress on the declarations of witnesses. Many appeal to their own experiences or to the witness of others whom they think trustworthy. The word "experience" must be taken literally, in an eyewitness sense. The apostle John begins his first letter with the words, "That which was from the beginning, which we have heard, which we have seen with our eyes, which we have looked upon and touched with our hands, concerning the word of life - that we proclaim to you." He calls on at least three of the senses for help. And the apostle Peter says somewhere, "We did not follow cleverly devised myths but we were eyewitnesses of his majesty." And elsewhere he remarks, "He appeared to us who were chosen by God as witnesses, who ate and drank with him after he rose from the dead." And John tells us somewhere about what they ate: baked fish. The apostle Paul and the evangelist Luke also appeal to many witnesses, and Luke makes another remarkable association in the dedication of his Gospel: He writes, he says, what had been handed down to him by those who from the beginning were eyewitnesses and servants of the word.

The chapter which deals with these things at greatest length is Hebrews 11, which begins with the sentence, "Now faith is the assurance of things hoped for, the conviction of things not seen." The rest of the chapter is then filled with a long list of witnesses of faith from the history of Israel: the patriarchs, Moses, David, and many others, who came into contact with the God of Israel in their lives and who bore witness to this experience. A short summary is given of each of these experiences. Now and then the summary is interrupted to describe what unites the witnesses: They were strangers on earth. They were in search of a better country. They sought the city of which God is the architect and builder.

In the history of Christian faith - to which I shall limit myself - things did not stop with the biblical evidence. The message spread over the world and was accepted by millions; for them it became an experience which they handed down as testimony to their children and those around them.

The witness of some of them, the "great" ones, has been preserved in the form of their words and a description of their lives, the story of their encounter with God. To answer our question, their testimony must be studied and weighed; that is the least one can do.

This is a subjective process. The degree to which a testimony makes an impression depends not only on its content but also on those who are open to it. Here is a far from exhaustive list of names of people who in this respect have been of great significance *for me*: John, Paul, Francis of Assisi, Blaise Pascal, Johann Sebastian Bach, Dietrich Bonhoeffer, Martin Luther King, Beyers Naudé, Mother Teresa.

The conclusion to be drawn from all the evidence can never be an incontrovertible proof that "God exists"; but for me the existence of God, his presence and effect in the lives of many people, is beyond any reasonable doubt.

It might be objected that the choice of witnesses seems arbitrary and even very selective. Why Francis of Assisi and not the Borghia popes? Why does the vote of Mother Teresa count and not that of the Inquisition? Over against Beyers Naudé stand the ideologists of apartheid, who derive their standpoint from the Bible; over against Martin Luther King stand the scandalous television evangelists who have made faith a commercial business. Isn't every vote in favor neutralized by a vote against? Isn't the total effect of twenty centuries of Christianity zero or even negative? What does the balance really look like when the weighing is carried out?

Recently, I was in the Dutch sculpture garden of the Kröller Muller Museum. There is a great variety of modern sculpture in this garden. At one point, I heard another visitor say, "The only thing that I am amazed at about these people [he meant the artists] is that they can sell such work at such a price." My question is whether, if one tries to fathom the significance of modern works of art, one must put the explanation of such a witness on the negative side of the balance. My answer is that one mustn't. Those who say such things only prove that they haven't seen anything. *Their votes don't count.*

A vote which does count - for me - is that of Pierre Janssen. This enthusiastic connoisseur goes around the country giving lectures with the aim of making modern art accessible to a broad public. I've been to some of the lectures. They are fascinating. With the utter dedication of his very enthusiastic personality, with a conviction which clearly comes from within (the way in which his whole body, his eyes, his expression, the gestures of his fine hands are involved), he tries to make his public share in his experiences with art. He doesn't prove anything. He doesn't prescribe what people must find attractive and what not, what people must see and feel. He tries to say what he himself sees and feels. At most he gives hints. He makes suggestions about how you yourself can seek and, with difficulty, find your way into this world of experience, how you can take part in it. That is a process which is never complete, for which a whole life is too short. But through such witnesses you get further, penetrate deeper into the meaning and significance of a world which at first sight you might pass by with a shrug of the shoulders.

The description I gave in the previous paragraph can be taken over almost literally if you ask, "How do I get through to the mysteries of the world of physics?", the question which scientists themselves face, or, "How do I share in the mystery of God?" In all cases, the witnesses are those who produce the evidence, who help people arrive at a deeper insight. The argument follows comparable paths. The witnesses must be critically tested, their evidence weighed; the wheat must be sifted from the weed. Selectivity is not only permissible but is a requisite. In every case it requires great effort. For what is worth taking trouble over is only achieved by taking trouble.

Science, art, and religion are all concerned with the riddle of human existence, the world around us, and the relations between these two. In contrast to what is often thought, science has no special place in the sense that it is "objective," whereas art and religion are "subjective." Even the scientist of the twentieth century must recognize that the observer and the observed are inseparable.

9 The Witnesses

"That which was from the beginning, which we have heard, which we have seen with our eyes, which we have looked upon and touched with our hands, concerning the word of life - that we proclaim to you" *1 John 1:1*

Science, art, and religion may all be occupied with the mysteries of the world and human existence, but on the scale of values of the people of our time, the three occupy very different places. Appreciation diminishes in the order I have indicated. Science may still enjoy a high regard, though recently a shift can be traced above all over its *consequences*, which are less full of blessings than had been hoped and expected. I expect that this disappointment will extend in the foreseeable future to the disclosure of the ultimate mysteries held out as a prospect by physicists like Simon Van der Meer and Stephen Hawking.

Art, and above all modern graphic art, has not until recently been experienced by the vast majority of human beings as a contribution to making existence meaningful. That too is changing: For an art exhibition of any importance, at present you have to stand in line for an hour before being admitted, to shuffle in dense rows past the works being exhibited. That may be a nuisance, but it is a good sign.

And religion? The churches are getting emptier and emptier, and that process has not yet come to an end. An expression like "Jesus is at the same time truly God and truly man" is irritably brushed aside as nonsense, as something people cannot cope with. A completely analogous statement like "An electron is at the same time a particle and a wave" is listened to with reverence and awe. Here it may perhaps be observed that church attendance or subscription to church dogmas is not the only, nor even perhaps the most important, barometer of human religious feeling, of our longing for the experience of God, for a meaning to existence from outside humanity. It is possible that the churches are just not successful in showing a viable way in a comprehensible language.

What about physicists? How do they regard the implications of their knowledge for a view of life? Generally speaking, I think it can be said that they are very cautious about this issue. In a recent article in the *Netherlands Journal of Physics*, entitled "Physics and philosophy do not go well together," some prominent Dutch physicists have been discussing this question. In connection with the books by Capra (*The Tao of Physics*) and Gary Zukav (*The Dancing Wu-Li Masters*), in which an attempt was made to connect quantum mechanics with Eastern and particularly the Buddhist and Taoist world-views, the physicist Casimir says, "I feel a certain aversion to attempts to combine physics with Eastern philosophies, and I am certainly not the only one." That is an understatement. The views of Capra and his like have generally met with chilly repudiation in the scientific world.

And rightly so, as has been argued that modern physics is bound up much more with the traditions of Western thought, and particularly Christian faith, than with Eastern philosophy. But apart from the question whether or not it is right to combine physics with any philosophy, many people feel that nothing of this kind must be attempted, that physicists must avoid it and go on doing their work, like old-fashioned road makers laying a road. This image comes from Gerard 't Hooft, a professor of theoretical physics:

> When a road maker lays a road he is not concerned why the road is being made and what other roads look like. He is concerned only with laying one stone properly before another according to the rules of his craft. So too the rules of quantum theory are fixed. Researchers know precisely what they must do in certain circumstances. That is enough. The discussion about philosophy, interpretation, is quite unimportant.

Many physicists nowadays maintain an equally restrained attitude towards religious views. In this respect they call themselves "agnostics"; in other words they say, "I don't know." It may be that there is a God, but then there may not be. The evidence is not enough to decide one way or the other. Believing colleagues are at least tolerated, although "respect" is perhaps too great a word. The subject is largely tabu. Religion is a private matter which stands, and needs to stand, outside the world of the daily experience of the physicist. The aggression with which a Simon Van der Meer declares his believing colleagues to be schizophrenics is more the exception than the rule.

It is voices like these with considerable volume which penetrate to the outside world, giving the person in the street the impression that belief in God is outdated and has already long been superseded by science. A minister tells me that he has already been informed by fifteen-year-old youngsters in Bible School that belief in God is an outdated affair; they said it with the triumphant faces of people who are conveying an irrefutable fact. I read recently that a professor of theology declared that he could not believe in the physical resurrection of Christ because he had never seen anyone come in through a closed door. Does such a person perhaps think that by making such a stupid remark he is gaining respect in the eyes of modern science?

I too am agnostic, not so much about religion, but about other matters like the existence of flying saucers, the workings of acupuncture, or astrology. On the basis of the evidence which I have on these subjects, I have considerable skepticism. But I would immediately add that this evidence need not mean much. I read something or hear something in bits and pieces, but I have never made a serious study and then weighed things up. The reason is obvious: *I am an agnostic out of lack of interest.*

Could it be that agnosticism about the question "Does God exist? Who is he and what is he doing?" runs along similar lines? In an earlier chapter, in Pascal's *Pensées*, we came across ideas which in the first instance point in the direction of agnosticism. Here, again, is one of them.

Nature presents to me nothing which is not a matter of doubt and concern. If I saw nothing there which revealed a Divinity, I would come to a negative conclusion; if I saw everywhere the signs of a Creator, I would remain peacefully in faith. But seeing too much to deny and too little to be sure, I am in a state to be pitied.

But he adds: "My heart strives to know and to follow the truth." That is the cardinal point. We have seen how Blaise Pascal sought and found and followed the truth in his life. We can certainly say that he wrestled with it.

Earlier in this book, I argued that anyone looking for an answer to the question "Is there a God? Who is he and what is he doing?" would do well to be open to the evidence of witnesses from today and yesterday who have spoken about their experiences of God. The witnesses must be selected and their evidence must be critically weighed. That is a personal, subjective matter, but understanding does play a role. It must be possible - also - to have a rational discussion about the selection and the process of evaluation. But there is more to it. "Being open" has a dimension in which the whole personality is involved; the evidence makes an appeal to both understanding and heart.

Helène Nolthenius and Julien Green have both written splendid biographies of Francis of Assisi. The one by Nolthenius is the more detached, the more critical, when it comes to testing the evidence by a critical examination of the sources in which the events are attested. Much then becomes doubtful, but even for Nolthenius a good deal of the evidence provided by this life stands the test. Both authors have explained how, from their youth, they were gripped by the figure of Francis and how this fascination has lasted a lifetime. For Nolthenius this one witness was enough to make her embrace the Christian faith in her youth and enter the Roman Catholic church. Later she left the church, by the way, which just shows that the persuasiveness of the evidence in matters of faith is never definitive and watertight - in contrast to Pythagoras' theorem, which has been demonstrated once and for all, and the wave character of light, which has been demonstrated beyond doubt. The argument is a lifelong process in which new evidence constantly comes to light and the old is provided with other weighty factors. The witnesses can put someone on the way, show a direction, perhaps convince temporarily, but we have to tread the way ourselves, test the pointers in our own experience. That can have different results, as is illustrated by Nolthenius and Green, who, though both impressed by the same witness, went different ways.

The saints were excluded from the Protestant milieu in which I grew up. The crucial factor was that the saints were venerated with a (perhaps almost) divine reverence which was not their due because they too were only human. That smacked of idolatry and had to be rejected firmly. The baby (the saints) was thrown out with the bath water (veneration of the saints). That spared us the sugar-coated lives of the saints which were related in Roman Catholic circles and which in this form were not very convincing, but it also robbed us of the witness of a long series of people who had heard the voice of God in their lives and had drawn the consequences. We had to accept Luther's *Sola fide, sola gratia, sola scriptura*: "by faith alone, by grace alone, by scripture alone," in which the word "alone," prefaced to three different things, at least makes a remarkable impression.

So it came about that it was only later in life that I was confronted with the existence of some saints and the content of their witness. Of these, Francis above all has become particularly dear to me. Why? Because of the peace, the love, the happiness that radiates from this man, over a gap of eight centuries. This was the happiness of a man who physically was increasingly wrecked with pain, the happiness of someone who lacked all those things which in our time are thought to lead to happiness: possessions, money, health, pleasure. But "lack" is the wrong word; he did not lack it, he rejected it with all the strength that was in him, almost fanatically. Lady Poverty was the condition of his happiness.

I shall not try to demonstrate in just a few lines what radiates from this man of Assisi. That is impossible. It has been done at length and in an admirable way by the writers I have mentioned and by others. The evidence is clear for those with eyes to see, though sometimes a good deal of rubble has to be removed to bring it out. I have looked for Francis' footsteps in Assisi and its surroundings, in the wide valley of Spoleto in the splendid Umbrian countryside. A few miles from Assisi lies Portiuncula, the moving, shabby chapel where the saint joined up with his first followers and where he also died.

It is still there. That the façade is embellished with a picture from a later period which is not without merit but misplaced is not so important. Much worse is the fact that around the little chapel, a gigantic, monstrous basilica has been built in which literally everything cries out against, clashes with, all that Francis stood for. It calls for considerable powers of abstraction in this baroque carnival tent to concentrate on the tiny little chapel, which is virtually swallowed up in it. If you try hard you can imagine them: Francis, Brother Leo, Brother Egidio, Brother Ruffino.

It is easier to picture them in the little monastery of San Damiano, where you can still see the tiny little roof garden in which the saint, physically already almost wasted away, composed his radiant hymn to the sun. I rediscovered him in the little hermitages and rock caves of the tranquil valley of Rieti, in Fonte Colombo, where he set down the rule for his order, in Greccio where he celebrated his famous Christmas feast, and in the hermitage of Poggio Bustone, where on a stone in the wall is written the greeting with which he saluted his fellow men and women, "Buon giorno, buone gente," "Good day, good people." Not because he thought that the people were so good but because he hoped that they would become as good as he trusted them to be in anticipation. I have seen something of him in his ragged habit which has been preserved, in his down-at-heel sandals, in the moving letter to Brother Leo.

It can happen like that: Those who steep themselves in the life of the great witnesses, in the words and actions of others who by their own confession have heard the voice of God and followed it, fall under the influence of the way indicated by these witnesses, and move along it. By way of example, there is the life-story of Helène Nolthenius. But usually it is not like that. The distance is great, in space and time. A good deal of material has to be rejected, and much is asked of our imagination. Far more direct is the impression of the witness of living people whom one knows closely: parents, friends, teachers. Their approval (or the absence of it, their disapproval) is more penetrating, more decisive, perhaps above all because it is also heard in one's youth, at the time when one is still open to everything.

In Holland, the Calvinist milieu of the 1930s-1950s in which I grew up has been made famous - or perhaps notorious is a better word - by the work of novelists like Jan Wolkers and Maarten 't Hart, who also grew up in this milieu and afterwards turned their backs on it. In their books they look back on it, either with bitterness and fierce repudiation (Wolkers) or with a somewhat more detached but no less devastating irony ('t Hart). I recall above all the description by 't Hart, often down to the smallest details: the narrow-mindedness, the poverty, the intolerance. The rationalistic way in which the text of the Bible was dealt with, being chopped up into fragments which were used as pieces of a dogmatic jigsaw puzzle that had to fit together precisely, or, even worse, were hurled at the heads of others. "I drove the minister into a corner with the Bible in my hand," I once heard someone say. The Bible is seen as an offensive weapon with which to threaten an opponent, so that he shrinks back in anxiety. A stream of Bible texts is used as a hail of bullets. This is a caricature, a perversion of what can be found in the Bible by anyone who can read a little.

This perversion was always there. If that had been all, I would have rejected it with loathing, like 't Hart, Wolkers, and so many others, because a human being worthy of the name has no other choice. But there was more to it. 't Hart's description is correct down to the smallest details, but one thing cannot be found there, the main thing. It is this: In these surroundings people could be found - and not so few of them - whom I could recognize as children of God. They conversed with God, received his words, did his deeds. In them I could recognize the rabbi of Nazareth. As the Bible put it, they were those who "walk as children of the light." Through them I kept on believing.

To keep on believing was less simple than such a sentence might perhaps suggest. The voice of these witnesses is not very loud; their lives are often not very striking. One can find little trace of them in publicity. They belong predominantly among those meek of whom Jesus says that they will inherit the earth, but whose legacy is bestowed at the earliest when the present owners are dead, so we may have to wait a long time. The meek are often drowned by those other voices which Maarten 't Hart has brought out so vividly.

It is very simple to write down, as I did earlier, that the latters' witness does not count because they clearly have not known God and thus do not know what they are talking about. But in practice, distinguishing between the "good" witness that counts and the "false" witness that can be denied causes problems, because the distinction between the "good" and the "false" witness which has been drawn hitherto is too simplistic, too black and white. It is far from being the case that here we have two kinds of people who are quite distinct and therefore can easily be distinguished. They are often combined in one person. So I have experienced that a person who at a distance has been a very important witness for me has on closer acquaintance seemed somewhat vain, full of himself, someone who did not so much carry on conversations as monologues.

I therefore have great difficulty with the sort of Christians introduced by Maarten 't Hart. Take the pair of elders from *A Flight of Curlews* paying a last visit to the terminally-ill mother of the main character (Maarten). It's impossible. They have piggy eyes, bottle noses; they are devoid of any human feeling; they're hard as nails. They vomit out biblical texts as sinister threats; they get at the mother and son wherever they can; they even abuse prayer as a pointed weapon. When finally the son kicks the "brothers" out, he does so with the full consent of the reader. The scene may certainly be said to be a fine example of narrative art, but it is no more than an extreme caricature. I have met dozens of elders in my life, and not one of them was represented anything like fairly by the description of the specimens which I have cited. However, I have certainly encountered some of the *elements* which can be found in 't Hart's description.

The same problem occurs with some descriptions of lives of saints. These become so exalted, spotless, and superior that they lack credibility. One can even have hesitations about a figure like Francis of Assisi. Thus his almost morbid antipathy to the female sex (with the exception of Lady Poverty and Clare) puts me off.

In the end it amounts to this. The witnesses that one encounters are not black devils or snow-white angels to whom one can give zeros and ones in an evaluation of their witness as a computer does. They are *people* who represent the whole spectrum of greys between black and white. The whole of their witness has an effect on the heart and the understanding, and guides an inner process of assessment, the results of which define the direction one takes. Sometimes one can sense or suppose - after the event - what witnesses have been of decisive significance. I shall end this chapter with a brief description of three people who I think have had this function for me. The most important of them was my father. I am doing this not in order to convince anyone, but merely as an illustration, by the hand of an illustrator who has only a little proficiency with this crayon.

MASTER BERGSMA

Master (as he was then still called) Bergsma was head of a Christian school in Delft. He was the headmaster, a very erect man with a moustache and a strict look. He *was* strict when necessary. He could also be gentle. My feelings towards him were comparable to what Christians call "the fear of God": love mixed with awe. He was also fair. Twice he expelled me from school for a week for bad behavior. There was no denying that it was deserved. When we left

school, he shook hands with all of us. He knew that I was
the only one in the class to be going on to high school. That
was aiming high in that environment. Too high perhaps?
The danger of pride was lurking. "Will you remain un-
spoiled, young man," he said, and almost crushed my timid
hand.

There is one day in that school year of 1943-1944 which I
shall never forget. One morning as we entered the class-
room, we found him behind the desk, slightly bowed. Tears
were streaming down his face. Shy, as though struck by
lightning, we sat at our desks and looked at what had happened there as though it was some bewildering natural
phenomenon which we would have thought impossible.
After a while he began to speak, and in fits and starts described what had happened. His oldest son, Folkert, had
taken part in the resistance movement against the hated
occupying forces; he had been arrested by the Germans and
shot the day before. The boy had done what he had to do;
there was no other way. His father was proud of that.

Then he turned his back, moved to the blackboard, and
took a piece of chalk. "Write this down in your books," he
said. With a firm hand, in the beautiful calligraphy that we
were used to from him, he wrote:

> Jesus be our guide through this world,
> and following in your steps,
> we shall go bravely with you.
> Lead us by your hand
> to our homeland.

There were three more couplets. Under it he wrote, "The
favorite hymn of Folkert Bergsma, shot by the occupying
forces." "Now we're going to sing it together," he said.

DR. CEBUS CORNELIS DE BRUIN

At the Rotterdam high school to which I went next, the
teaching in Dutch language and literature was given by

Dr. C.C. de Bruin. His pupils nicknamed him "Zebedee" and I only understood the origin of this when I saw his full Christian names for the first time in the notice of his death. In a sense, the heavy sound of the nickname described him well. He was a heavily built, broad man, with a big wide head and a full voice, which resounded through the school with enormous volume, particularly when he was angry (once or twice a year).

He was an extraordinarily learned man, whose lessons were wasted on many of his pupils, but valued all the more by others. Later he achieved the status he deserved: professor at Leyden University. He was awarded the very rare prize of Master of the Dutch Society for Literature. Twenty-five years after I left school, I received, at my home address, an invitation to attend Professor de Bruin's farewell lecture at the University. I was utterly amazed. I hadn't seen him for so long; how had he discovered where I lived? For that reason alone I went. Nor was I the only one; there were hundreds of old pupils there. At the end of his brilliant lecture on Dante, he said: "I'm going to end my last lesson with a poem by Nicolaas Beets which all my old pupils know by heart. It emerged from silence as the expression of a religious-poetical experience and returns to a silence that speaks louder than words."

"The tops of the mulberry trees rustled"
God went by
No, not by, he lingered;
He knew what I needed
and spoke to me.

Spoke to me in the silence,
the silent night.
Thoughts which tormented me,
persecuted and disturbed me,
he drove away gently.

He let his peace dwell
on soul and senses.

I felt his fatherly arms
cherish and protect me
and fell asleep.

The morning which woke me
I greeted happily.
I had slept so gently
and you, my shield and weapon,
were still near.

What was marvelous about this event (quite apart from the mysterious detective work by which he had discovered all our addresses after twenty-five years, and about which he maintained a mysterious silence even at the reception)? The fact that he took it for granted that, though he had never made us learn this poem by heart, we would all nevertheless know it by heart. And that was the case. I did. What he meant to say was: Of course you've forgotten all those hundreds of lessons which I gave you over the years. But of course you've remembered the only thing I had to communicate, the only thing that really mattered. And so it was.

MY FATHER

I can still see him clearly as he sat there the evening we visited him in the hospital. That noon, the surgeon had been to see him and had told him the results of the investigation. He was incurable. He had only a few months to live. He told us and cried a little. Then he said: "I've been able to live seventy-two years and I've enjoyed every day of it. I was so glad still to be going on for a while." I knew that was true. There was no question here of a laborious existence being romanticized in the face of death into something that it had never been, though that could have been possible. He really had been happy.

What did such a happy life look like? At first sight, like this. He was the oldest son of a poor farming family with eleven children, and had a sharp mind. He had particularly wanted to study but never got the chance. After going through elementary school, he joined his father among the cows. Later he got engaged to a girl who immediately afterwards contracted tuberculosis. During the years it took for her to be cured, those around him almost unanimously urged him to break off the engagement. Such a wife could not grow into a healthy strong farmer's wife, the kind necessary for the job. He chose stubbornly to wait, finally married her, and began a business of his own: a herd of ten cows bought on borrowed money, a piece of rented land, the hired stalls. The burden cost him his last cent.

This was 1930, a year after the crash of 1929. The crisis was world-wide. The yield of the business was quite inadequate even for a minimal existence. He decided to go into the city to sell the milk from his cows, becoming a milkman and a farmer at the same time. He toiled through the city with his pedal cycle. One customer per street, murderous competition. He worked seventy-hour weeks and just managed to keep his head above water. In 1940 the war broke out. Now was a great chance at last to make money, because on the black market his produce was worth its weight in gold. He did not join in, and in the postwar currency reform not a black cent was required of him. So he remained poor, although things gradually got better after the war.

When he was fifty, he had a whole week's holiday for the first time in his life. Before that, a holiday consisted in cycling once or twice a year on a summer day to the beach (about ten miles away), he and mother each taking a child. In the morning he had already done the milking, between five and eight, and he had to be back in the afternoon for milking time. One of my classmates at the elementary school called that kind of family "people who take a day trip" because his parents rented a house at a nice beach for a whole month. We were a kind of refuse which polluted the beach.

What did my father do with his "spare time"? The most important part of this was on a Sunday, between eight and three (before eight and after three there was milking). Between eight and nine he took a nap. Then he dressed in his elder's get-up (from the age of twenty-nine he was an elder almost without interruption): black striped trousers, a black jacket and waistcoat, a stiff white collar which was fixed with two studs to a collarless shirt, and a grey tie. Thus equipped, he went with his family to church. There we saw him, just before the beginning of the service, entering the church in the line of elders and deacons, all wearing the same uniform. After church there was coffee, this one time with a cake, and dinner. Then sleep until three. He devoted weekday evenings, as far as one could still call these free time after the working day I have already described, to the work of an elder: visits, meetings, study groups. In addition, from a very early age, he was also active in running a farmers' cooperative, which with growing success opposed the exploitation of the small farmers by the capitalist milk factories.

These are a few of the most important "hard facts" which could provide enough material for a biography with a bitter tone, a slave's life. Nothing of what today we would call "fulfillment." He read few books. He never went to a concert, let alone played an instrument, and never saw inside a museum. How can someone like that truly declare at the end that he enjoyed every day? How is it possible that I would not want to have had any other father than him? It was because he loved God, God, with whom he had to do, whom he encountered in the Word, to whom he spoke in prayer and from whom he received an answer, whose light shone through him. That was what was given him to enjoy, day in and day out. The rest was secondary. Poverty? If you pray every day, "Give us today our daily bread," and get it, what more do you want? "My wife hasn't a fur coat but she has a pretty apron."

The Word was central. It was read out three times a day: at midday and evening meals and before going to bed. After the last reading, he and mother knelt on the floor, in front of their chairs, as we discovered when, around ten, we opened the door of the living room and quickly shut it again. They worked through the Word, from A to Z, but the book of Psalms had a very special place. There were special psalms for special occasions. If anyone in the family had a birthday, it was Psalm 103: "Praise the Lord, O my soul, and forget not all his benefits." On Liberation Day, May 5th, Psalm 124: "If the Lord had not been on our side, now may Israel say, when men rose up against us, they had swallowed us up quick." If there was a death in the neighborhood, Psalm 23: "Though I walk through the valley of the shadow of death, I will fear no evil, for you are with me." On New Year's Eve at quarter to twelve the radio would be switched off and we knew what was coming: Psalm 90: "Before the mountains were brought forth and ever the earth and the world were made, you are God from everlasting and world without end." After that he gave thanks for "all the blessings and benefits" which had come to us from God's hand during the past year and prayed that God would go with us into the New Year. Of course there was a psalm for New Year's Day, Psalm 91: "Who so dwells under the Most High, shall abide under the shadow of the Almighty."

That is what he lived by. This was expressed not so much in a good deal of pious talk (he had a marked dislike of cant) but in his attitude to his fellow human beings. "Those who have never seen their neighbors have never seen God," he sometimes said, and acted accordingly. On the eve of the winter of famine, in September 1944, when the food shortage was already beginning to bite, he sold almost all his potato harvest at a normal price to those who were in need. After that he gave so much away of the normal winter store which he had kept for his own family that this ran out too soon, so we sometimes had to eat bulbs and sugar beet, obtained on laborious bicycle rides which he went on with me to the bulb-fields.

On one such "famine trip," on my own initiative, I once sneaked a few lettuces. To my great anger, he gave them away to someone in more need than we were. He shared out the milk which he kept from the delivery unit controlled by the occupying forces, especially to people in extremities who were pointed out to him by the doctor. "Unless you help X," the doctor would say, "X will soon be dead," and then he would help X as much as he could.

All that, and much more, happened in a completely natural way, with no need of explanation. Was he perhaps a saint? Not in the sugary, unworldly sense of the word. His priorities in life were fixed. Here is one last example. Had he been asked, "If you had to choose whether your son should become a believing street cleaner or an unbelieving professor, which would you choose," then without a second's hesitation, he would have opted for the former. A believing professor would of course have been more attractive, but secondary. The most attractive prospect of all was that of having a reverend son, a servant of the Word.

That had been his own unfulfilled dream, but there could be no question of it in the farming family in which he was brought up. At the age of twelve, elementary school completed, the duty to learn fulfilled, there was only one possibility: putting your shoulder to the wheel, getting among the cows, helping to keep the family alive. Already then, he said, he had decided that things would be different with his children.

So he sent me to high school. His family told him that he was crazy. Couldn't he use this boy in his business? Of course he could. That would have turned a fellow eater into a fellow worker, a relief for a family which already hardly had enough to go around. He never gave in, obviously. It had to be a high school where they taught classical languages, which were an indispensable grounding for the study of theology. He made it clear that this was the supreme goal. It also went without saying that the school had to be a Christian school. Since there wasn't one in Delft, I had to travel to Rotterdam, adding the cost of a train season ticket to his expenses. I still know precisely what it cost: 9.60 guilders, a monthly pound of flesh.

So I knew what was expected of me, but along the way it became clear that I couldn't bring myself to do it. I felt no "calling," something which was thought necessary for becoming a minister in those days. Moreover, I had a slight preference for the exact subjects. At the end of the tenth grade, the choice had to be made, and I chose exact sciences. I still remember how hard it was for me to tell him, because I knew how deeply disappointed he would be; indeed one could put it more strongly and without exaggeration: His world would fall apart. I told him. I saw him swallow, and then he said, "You must take the direction for which your talents best fit you. Only you can be the judge of that." Later, when I told him what I wanted to study, he said clearly what he thought of it. "An engineer's no more than a carpenter with a top hat." A useful craftsman, like he was. There was nothing exalted about it.

This part of the story has a happy ending. He had sent my younger brother, who mended the bell at home when it got broken, and was keen on building radios, to the technical college. There - at best - he was to become an engineer. When he had gone through technical college, he told his father that he wanted to become a minister, and that is what happened. I think that nothing in my father's life made him happier.

After the fatal message from the surgeon, he wanted to go home, and there he spent his last months. In peace and quiet, he tidied out his desk, put his papers in order, and passed on information to those who needed it. Even then, pious language did not pass his lips. "They know it well enough," he must have thought.

We buried him on July 14th, 1976. July 14th, "le jour de gloire." At the funeral service, we sang a hymn, two lines of which sum up the essence of his life:

Those who are planted in God's house
blossom in God's light.

After the funeral, an old woman said to me, "Condolences and congratulations." It couldn't have been put more succinctly.

10 Two Ways

Search me, O God, and know my heart!
Try me and know my thoughts
And see if there be any wicked way in me,
and lead me in the way everlasting
Psalm 139:23-24

In an interview from which I quoted earlier, with the Oxford philosopher Leszek Kolakowski, formerly a Marxist, the following exchange takes place between the interviewer (I) and Kolakowski (K):

I: Did you become a Christian?
K: In a certain sense, you could call me a Christian.
I: In what sense?
K: I don't want to go into that.
I: Why not?
K: Because it isn't anyone's business.

That's plain speaking. The way in which I experience my belief, says Kolakowski, is a private matter which I don't want to make public. I'm not an exhibitionist.

In the previous chapters, I have done no more than indicate how I came on the way to faith. Things can't stop there. At this point, the reader is quite justified in asking a few questions, like "What happened to you on this way? And on the other, that of physics? Have these two anything to do with each other, or are they completely different?" To put it perhaps more strongly: Is it possible to take both ways together and still preserve one's integrity? Or is one "schizophrenic"?

I would most like to agree with Kolakowski that this is nobody's business, but I'm aware that this isn't possible. I can't avoid it. I must at least try to bring some clarity to the questions which have been raised, without lapsing into exhibitionism on the one hand and empty words on the other. I shall do so by going into the nature of the two ways in more detail than I have done so far. We shall then see that they have many characteristics in common. At the end we shall consider a number of points where they intersect.

THE WAY OF SCIENCE

How does anyone come to be a scientist? Perhaps in most cases, the best answer is still "by chance." Young people who towards the end of senior high school face the important decision of choosing a specialist study usually have extremely vague ideas about what such a study involves, let alone the professional activity to which it provides access. They study some illustrative material, look around the open days organized by colleges and universities for a few hours, and sometimes find what they see "very nice" and sometimes not. A teacher who can teach a subject in an inspiring way can be a great influence. One's father's profession, often the only one of which a child has a more than superficial idea, can also be an important factor. A remarkable number of doctors' children go on to study medicine. There's also whatever is fashionable. At the beginning of the period of space travel, in the 1960s, the aeronautics departments were full; in the 1970s, the period of a "livable society," this was the case with sociology and behavioral studies; in our time, it is the same with environmental studies. In the end, financial prospects do not play a major role. For most young people, the most important factor is whether they expect that the study they have chosen and the professional practice which follows can give them satisfaction. On the basis of these expectations (or even more vaguely, these feelings), they choose.

That is roughly what happened to me. At the end of senior high school, there were three subjects in which I was especially interested: Dutch language and literature, classical languages, and physics. They were all three taught by inspired teachers. Why did I decide on physics? First, because the two others would almost certainly have resulted in my becoming a teacher; I did not find that unacceptable, but it was too risky because of the lack of alternatives. Physics offered an alternative: being a researcher in a laboratory. I had hardly any idea what that involved (though it seemed all right to me), but at least it was a possible choice. Secondly, the atomic age had just begun and was exercising an enormous influence. That this beginning was marked by the two devastating flashes of light over Hiroshima and Nagasaki could not temper the euphoria. Anyone who asked a first-year physicist at that time "What is your special interest?" usually got the same answer: "Nuclear physics," penetrating the deepest secrets of the atom. Thirdly, physics could be studied in Delft, where we lived. That cut down the costs, which was an important factor. Granted, the emphasis there was on applied physics, but - as I thought at the time, though on what basis, I'm not sure - that would not make much difference. And finally, my father's profession, that of a farmer, was the only one that I knew at close hand. It is (was?) splendid work, in direct contact with nature and animals, with a tangible and meaningful product, but the prospect of seven days a week of ten-hour days was not encouraging, particularly since the material reward led to a very sparse existence. My father certainly did not try to make me follow in his footsteps: quite the contrary. He used this possibility as a goad: "Just remember, if your studies don't work out, I'll hire you as a farm worker."

So I went to Delft to study physics, with all that behind me and a very vague future in prospect. In the first chapter, I described the spiritual shock which that gave me. The change of climate was radical. The package of subjects in the kind of senior high school I went to (gymnasium) is varied and harmonious. It speaks to all levels of the personality. After an hour of physics, someone comes to read Homer with you or to interpret a poem; you relax your muscles at gymnastics for an hour or you learn to listen to music. At the technical university of Delft, it was measuring, counting, calculating all day long - a plunge into a bath of cold water. It took me years to find a balance, and in fact that is still going on. Novels, poems, music had to provide it, and did. Shakespeare, Bach, Mozart, Brahms and many others had to sort out the balance which had been upset by Newton and others. And then there was still God.

Instruction in physics, like that given to us by our professors, initially had a strictly *dogmatic* character - strange though that may sound. *Truths* were proclaimed, natural laws formulated in mathematical language, which up to that moment had withstood the test of time, which had withstood any critical testing by experiment. All the physical phenomena observed up to that moment could be described by them (at least in principle). Some of them we could also verify ourselves. Our teachers thought of questions to which they knew the answer and which we had to be able to solve on the basis of the theory we had learned. During practical work, we ourselves could carry out experiments, or at least go through procedures which were largely settled and which with careful use produced the results which hundreds of students before us had already produced. These results, both those of the exercises and those of the experiments, were always in accord with existing theory (if that was not the case, we had made a mistake and had to repeat the experiment). So there developed among us the solid conviction that what was taught us was *true. It was true because it worked.*

But as we went on, it was made increasingly clear to us that this was not the main aim of our study. The aim was not that when we got to the end of the study, we should say, "I believe that it is all true, I cannot find any fault with it." The aim was rather that we should master existing theory and experimentation thoroughly in order later, *on the basis of this,* to be able to work independently, as fully-fledged physicists. It was that we should be able to make up exercises to which no one knew the answer, be able to make up and carry out experiments which no one had ever done before. It was - and perhaps this was the most difficult thing of all - that we should learn to formulate questions to which a meaningful answer could be expected, an answer which would contribute to deepening insight. In short, it was that we should pioneer new ways, but ways which joined up to all that our predecessors had achieved. We were to become road makers who could add our own highly personal stones to a road which had been laid down over the ages by all our predecessors, the great and the less great. The great, of the status of Newton and Einstein, had built bridges at places where it looked as if the way had to end on the edge of a ravine. After that the less great could go on toiling in land they could build on.

So our study served to teach us how to master the "truths," the foundations which were disclosed to us at college, in the sense that they began to live and become operational for us. We learned to use the equipment: the screwdriver and the soldering iron, the camera and the microscope, electronic instruments, and above all mathematics. So the day approached when we were declared to be fully-fledged physicists. The step which preceded this was the postgraduate project. For the first time, by the professor whom we had chosen as supervisor, we were given a question to which no one knew the answer, not even the professor. We then had to wrestle with it for a year, and at the end set down our findings in a report. The conclusions of such a report seldom contain the definitive solution of the problem raised. They usually describe how one has got a little further in the direction of an answer. They contain suggestions about what must be done next. They raise new questions, to be answered by successors. Sometimes they simply report that the question raised leads to a dead end and that other questions need to be asked. That too is a positive result. No new stone has been laid on the road, but at least those who come after us have been shown a direction in which it is better not to make attempts.

The completion of the post-graduate project, the proof of competence, leads to the awarding of a doctoral diploma, the end of study, the declaration that one is grown up. That could hardly be put more unfortunately. Grown up? Fully grown? Growth has yet to begin. A seed has been planted, no more. The diploma is no more than a statement by our teachers that they are now prepared to let us go forward. *From now on, we may join in.* We are admitted to the community of scientists, which is international, and of all times.

This community is essential. Those who begin research in physics are not cast into no man's land, the middle of nowhere. They stand at the provisional end of an ongoing way and can go on further from there. They stand on the shoulders of those who have preceded them. They put themselves firmly on the level that their fellow scientists have reached all over the world, on the same front, and try to get further. They study the testimonies to the experiences of other scientists which are to be found in the scientific literature. As we saw earlier, these experiences can in principle be verified, repeated. That may be the case in principle, but it is seldom so in practice. We saw earlier why not: The repetition of other people's experiments is usually beyond our reach. We do not have the means, the skill, and the time. And there is yet another reason. It is not worth it. Almost always, we would discover what the other person discovered. Sometimes someone is caught out in a mistake; it is even rarer for a deliberate fraud to be unmasked. So the possibility of repetition is usually put into practice only if there are justified reasons for mistrust. For the rest, we accept the witness of our professional colleagues as trustworthy, if only because in fact we have no choice. That trust is not often put to shame. The community of physicists seems to be a collection of largely serious and careful people of integrity. In their midst, in collaboration with them but also (as we saw earlier) in competition with them, you go your own way.

Gradually you get to know your colleagues. You meet them in their work, the specialist literature that is published on a subject; you know their names and meet them at international conferences on a particular topic. Names take on faces, the great names first. They distinguish themselves at the meetings as they distinguish themselves in their work. There are the draught horses, the inspirers, those who give the main lectures and dominate the discussions. In the corridors, they are constantly surrounded by the troops who want to talk to them and work with them. If you're lucky, you can book a place at their table in advance to take part in a "working breakfast." They are also about in the evening, when, exhausted from a day on their feet, people seek diversion in bars and cinemas. In addition to an exceptional aptitude, something else marks them out which is just as important: *total dedication to physics*. In this way - sometimes - they reach the absolute top: the award of the Nobel prize, beatification for scientists.

There are different kinds of physicists. The subject has grown so huge that no one can see it as a whole any longer. The community of physicists is split up into sub-groups. One is a nuclear physicist, an astrophysicist, a solid-state physicist, and so on. Within each sub-area people speak a language of their own which is difficult for outsiders to understand. The interface between these sub-areas is small, and for much of the time, people just live side by side. It is difficult to tell other people what one is involved in. Only the great leaps forward, like quantum mechanics, make their influence felt in almost all the sub-areas.

There is yet one more sub-division into kinds of physicists, this time only into two, a division which runs through all the sub-groups: There are experimental and theoretical physicists, the doers and the thinkers. There are experimenters who construct apparatus to bring new phenomena to light and to detect new characteristics of matter. And there are theoreticians who, armed with paper and pen, and in modern times with computers, reflect on these phenomena and try to bring them together in as comprehensive a theoretical framework as possible. If things go well, these two form a harmonious, perfect duality in unity. The theoretician explains and predicts; the experimenter measures and tests. We have already come across an instance of this: The theoreticians Steven Weinberg and Abdus Salam predicted from behind their desks the existence and the properties of W and Z particles; Simon Van der Meer and his colleagues confirmed that in their experiments. Things don't always work out so harmoniously. Sometimes the experimenters accuse the theoreticians of being too ethereal, going around too much with their heads in the clouds and being too little use in devising meaningful experiments that can be carried out.

Finally, something about the nature of research in physics. I shall not take your time with an account of my own research, but simply mention three characteristics of the work as I have experienced it. It is *unique*, it is *exciting* and fascinating, and it is *no fun*.

The work of the researcher in physics is essentially different from that of, for example, the baker. The baker bakes today the same bread that he baked yesterday, and he will do the same thing again tomorrow. Moreover, this bread is not substantially different from that produced by other bakers. If the researcher has chosen a problem well, he or she is the only person in the world occupied in this way. Never before in history has anyone done the same thing. It may well be that at the same time other people may be occupied with the same problem elsewhere in the world, with or without his or her knowledge. What counts is who comes out first. The result is then unique. Once the researcher has published a result, it is done for ever: It need never be repeated by anyone. It is his or her personal contribution, with his or her name on it. Those who come afterwards can refer back to it.

The work is exciting. The best comparison is still that with a detective in a complicated crime novel. It is the collection of bits of evidence which are hidden among piles of irrelevant information. It's like looking for grains of gold in the sand. You overlook them, you constantly get off on the wrong foot. Apart from logical combination and deduction, it also calls for patience, intuition, creativity, until at some moments the pieces fall into place and form a logically coherent whole. These moments are not numerous, but satisfaction is too weak a word to express such an experience. Joy, happiness, comes closer to it.

The way which leads to such moments is fascinating, but it is no fun. Fun is the word which describes the world conjured up for us in advertising campaigns. There we see only cloudless skies, radiant faces, happy families. There isn't a drop of sweat to be seen, and no tears are shed. Here is happiness that costs nothing. The scientific researcher doesn't live in such a world. Edison said of his work that it was 99% perspiration and 1% inspiration. It is hard labor, desperate wondering whether you may not be going up a dead end, anger about trivial matters like the electricity or the water being cut off, ruining the work of weeks. All that is forgotten in the greatest moments of breakthrough. Or rather, it is precisely against the background of the toil that those moments get their real splendor. It was worth the trouble above all, because it involved trouble.

Was it truly worth the trouble? By what criteria can that really be measured? A few critical questions about the description I have just given of the happiness that is experienced in scientific research may not be out of place here. Isn't it totally "I-directed"? Doesn't it seem suspiciously like self-satisfaction? Isn't it just the search for a pleasant feeling, perhaps on a rather higher plane than that on which the needs of the present-day consumer are satisfied, but essentially the same? Don't pride and vanity play a great part here? Don't I feel myself superior to the baker, because my work is so unique and his is so everyday? Why should society spend so much money on science? What about the social dimensions of the work? Surely it unmistakably has social consequences? What does one do it for? Does it make sense? At this point a couple of lines of a poem by Jacques Bloem, a Dutch poet many of whose verses are written on my heart, occur to me.

> Is it enough, one or two poems,
> to justify an existence?

I suppose that if Jacques Bloem had read the description of the nature of scientific research given above, he would have recognized it. Making poems is not a matter of waiting for inspiration and then, when it comes, writing down what it dictates. Inspiration, ideas are indispensable, but the way to the completed poem in which every word is in place can be long and laborious. Given the sublime quality of his poems, Bloem must have known that feeling of happiness many times: That's it, that's good. And yet he still asks whether it is enough to justify an existence.

The question of the justification of our existence is a *religious* question. In a previous chapter (Chapter 6), I quoted Blaise Pascal, and I repeat the quotation here in an abbreviated form:

> We do not know who has put us here, what we have come to do, what will become of us at death. I see other persons around me of a like nature. And thereupon, I wonder how people in a condition so wretched do not fall into despair. I ask them if they are better informed than I am. They tell me that they are not. And thereupon, these wretched and lost persons, having looked around them, and seen some pleasing objects, have given and attached themselves to them. *I have not been able to attach myself to them,* and, considering how strongly it appears that there is something else than what I see, I have examined whether this God has not left some signs of himself.

There are plenty of opportunities for diversion from a few pleasing objects. The commercial world offers them by the handful. But also a stage higher than mere consumerism, at the level of science on which Pascal moved, with the satisfaction, the triumphs, and the feeling of happiness which he doubtless must have known, he remarks that "I have not been able to attach myself to them." No justification for an existence. Can anything else be found? Does God make himself heard in any way?

It's time to say something more about that. But I should warn you in advance that I shall not be giving any clear answers to the questions I have raised, any neat solutions, any formulas for the meaning of existence. At most the contours will point to a way one can take, and which I took in a tentative and exploratory way.

THE WAY OF FAITH

The answer to the question whether God exists need not in itself be particularly interesting. For example, it was hardly by coincidence that in deism a way of regarding God arose with the mechanistic world view after Newton, with the beginning of the Enlightenment, according to which God is the mechanic who constructed the world in the beginning, set it in motion and then left it to run by itself - just as a clock maker makes a watch, winds it up, and leaves it to run.

Whether or not such a God exists leaves me cold. The only possibility that is of importance is whether God exists in such a way that he is involved in the world and my existence, that he "does not let go what his hand began," that he has shown a way through life which can be taken and which comes out somewhere. Even that need not be particularly appealing. Malcolm Muggeridge put it in the following, somewhat challenging, way:

> Is there a God? I myself should be very happy to answer with an emphatic negative. Temperamentally, it would suit me well enough to settle for what this world offers, and to write off as wishful thinking, or just the self-importance of the human species, any notion of a divine purpose and a divinity to entertain and execute it. The earth's sound and smells and colours are very sweet; human love brings golden hours; the mind at work earns delight. I have never wanted a God, or feared a God, or felt under any necessity to invent one. Unfortunately, I am driven to the conclusion that God needs me.

God needs me for something, he wants me, says Muggeridge, and that is not very agreeable in itself. It is even disquieting. I would rather be left in peace, so that I can devote myself to my pleasures. That would be far less trouble. But I cannot avoid God.

In the previous chapter, I indicated how I came upon the way of faith. Through their teaching and example, the witnesses from my youth, parents and teachers, aroused the suspicion and a gradually growing certainty that God not only exists, but that he is also at work in human lives.

The instruction in the faith that we received also had a dogmatic character. In it *truths* were proclaimed. Let me give a short impression of this. God created the world and human beings. After creation he rested on the seventh day, considered his work, and concluded that it was very good. Human beings were appointed to rule creation and develop further, but that was not enough for them. They wanted more, they wanted to "be like God." That was the "fall," which drove them from the paradise in which they lived in harmony with God and his creation. One of the first things that fallen man does is to kill his brother. But God does not turn away from him. He appears in the life of Abraham, the patriarch of the people of Israel, to whom he assigned a land to live in. Through Moses he gave the Torah, the Law, the rule of life as he intended it. He pointed out the way and continued to do so, through the voice of his prophets, even when the people kept departing from this way. In the end he sent his Son, in whom he definitively shows who he is, Jesus, who says of himself that he is the way, the truth, and the life. The one who caused scandal by this claim was nailed to the cross, but after three days he rose from the grave.

He lives, even now, and gives life and liberation to those who believe in him and want to follow his way. He is himself the guarantee for the final consummation of the world, the coming of the kingdom of God, briefly described by one of the apostles as "a new heaven and a new earth in which righteousness dwells."

That, in an extremely brief summary, is what was presented to us as "the history of salvation" by those who brought us up. It was further supplemented by all kinds of statements which the church had formulated in the course of centuries in order to make the Christian faith more precise: dogmas, like those of the twofold nature of Christ, the divine tri-unity, and so on. Truths: And now people like Paul Davies (Chapter 7) think that a believer is someone who subscribes to a whole series of such "truths" and regards a number of historical facts as correct, if need be against all sound understanding and all the scientific evidence. For him, believers are those who as it were sign a document on which the facts and truths are summed up, "Agreed, the date...," and rigidly keep to this declaration all their lives. But things aren't like that at all, if only because it would be completely incomprehensible why something of that kind should happen century after century, from one generation to another.

A believer is someone who has taken a way, just as a physicist is someone who follows a way. Because the two ways have so much in common, I shall repeat here a brief description of the latter way. We arrived at it by the inspiring instruction of our teachers. They disclosed to us the fundamentals, the "truths" of the discipline. And they did more. They showed us how these truths could be made operational. They gave us the equipment and taught us how to use it. Only in experimentation, in our own experience, did the principles come to life for us. We could be involved in them ourselves. The only meaningful way of doing that is in connection with the physicists of the present and the past. We might join on the basis laid down by our predecessors, along with our contemporaries. It is a way with its ups and downs, toil and sweat, but also with moments of deep satisfaction. On the way you become wiser but also sadder. Insight is deepened, but it leads to a growing sense of one's own limitations and inadequacy; the more I know, the more I discover that I know nothing. It makes for modesty and humility.

That is roughly what the way of believers looks like. At the beginning stand the witnesses from their youth, parents and teachers. They gave us the basis. They showed us the equipment by introducing us to what Pascal calls "the custom": They taught us to pray, to read the Bible, to sing, and took us to church services. To begin with, we felt that these were endless, but gradually it began to dawn on us that something was *happening* there: A way was being shown, and prospects were being opened, words were spoken there which touched the heart, words from the other side. And the most important thing was that in this daily converse with the witnesses, we could see that these words also *worked*. They did not just remain words spoken by pious people (who were in fact sparing with them) but they were *experienced*. So it slowly became clear that we also wanted that. Not because it was so attractive, or so easy, but because of the light that shone there, the tranquillity which radiated from it, the trust. Because a purpose was evident there.

There then comes a moment when you say that out loud. In Christian churches it is called "making a confession of faith." It takes place in church services, particularly at confirmation services on Easter Eve. There a group of young people say yes to the questions put to them, which amount to whether they are ready to take God's way. The only meaningful way of doing that is in connection with believers of the present and the past. So that "yes" is said in the midst of the community. Like a doctoral examination in physics, it is a beginning. The seed is sown, the direction is given. From now on, you may join in.

What does the way of God, which at the same time is the way to God, look like? Where is it to be found? A good description of it is perhaps the following verse, which is a version of Psalm 85:

> Wherever he goes peace goes before him,
> love and faithfulness follow in his steps.
> Justice goes before his face,
> it blossoms wherever he sets his foot.

This is about the footsteps, the traces of God. It is a description. Where God is, he is at work, and the visible signs of his presence are peace, justice, love, faithfulness. Where these are found, there God is. Anyone who shares in these works shares in him, follows in his footsteps. In the Sermon on the Mount, Jesus says something like: "Blessed (i.e.

full of God) are the peacemakers, the merciful; blessed are those who hunger and thirst after justice." No "truths" are proclaimed here; what this is about is *doing* the truth - a unique biblical expression. "Whoever does the truth goes towards the light," says Jesus somewhere.

The chapter of Hebrews which I quoted earlier, in which a long list of witnesses of faith from the history of Israel is reviewed, ends with the following conclusion: "Therefore, since we are surrounded by so great a cloud of witnesses, let us run with perseverance the race that is set before us, looking to Jesus." That is the essence of "the way": It is as simple as following Jesus.

But that's not so simple. On this way all the signposts are turned round 180 degrees from those of the world in which we live. There money is thought to be the key which opens all doors. Here it is easier for a camel to go through the eye of a needle than for a rich man to enter the kingdom of God. The world is concerned with reputation, influence, power. On the way we hear, "Those who exalt themselves shall be humbled and those who humble themselves shall be exalted." "If I am weak, then I am strong," and so on. Jesus sums it up in one sentence: "Whoever will save his life shall lose it, and whoever will lose it for my sake shall find it." And the life of Jesus was summed up by an apostle in one word: He *emptied* himself. He gave himself away to the death which followed. It is really no wonder that the writer of the letter to the Hebrews calls those who want to follow this way "strangers upon earth."

I wrote that physicists become increasingly more aware of their limitations. That makes for modesty. Something of the same kind happens on the way to faith, but at a deeper level. Thomas à Kempis, the author of *The Imitation of Christ*, puts it like this:

> As you meditate on the life of Christ, you should grieve that you have not tried more earnestly to conform your-self to him, although you have been a long while in the way of God.

Those who look into this mirror have no reason to be content with their Christian achievements; indeed - since here too the signposts have been turned round 180 degrees - the greater one's "achievements," the less content one can be. What Thomas, who was a very holy man, found, we can also find in the lives of all great saints: an unfathomably deep repentance over their own failings in discipleship, coupled with great penitence. The closer to the light one is, the deeper the sense of one's own darkness. The way of discipleship is the least suited of all ways to give people the feeling that they can prove themselves by their achievements, that they can justify their own existence. But - and here is the paradox - that does not lead to despair. For on the way they discover, to their relief, that they do not need to justify themselves *because they are justified.* I meant something like that when I said of physicists that the foundations of their science come to life for them in experience to the degree that they join in that experience. The Christian doctrine that "Human beings are guilty before God of failures in discipleship. Guilt is forgiven them to the degree that they repent" is not primarily a theoretical statement to which one can subscribe or which one can reject, but a truth which becomes true to the degree to which it is experienced. That is what the Bible calls "walking in the truth."

That's just one example. I shall not add others because it is not my intention here to provide an account of my spiritual experiences (which are in any case meager, and there are enough people from whom one can learn more in this area). But I shall briefly indicate some other parallels between the two ways.

In the first instance, faith may be a strictly personal matter, but it is experienced in fellowship with believers of all countries and all times. One is not left in no man's land, but goes along the way with others. That is not always simple, because these others, like oneself, are people to whom nothing human is alien. An extra difficulty in this sphere is that not everyone avoids the temptation to endorse their own views with an appeal to divine authority. The Crusades were not the first or last time when the cry

"It's God's will" had such disastrous consequences. On the first page of his book, Thomas à Kempis gives the following warning: "Of what use is it to discourse learnedly on the Trinity, if you lack humility and therefore displease the Trinity?" Too little attention is paid to such words. That does not alter the fact that one does not have to manage on one's own but is directed towards the community.

There are different kinds of believers. There are Roman Catholics, Episcopalians, Congregationalists, Lutherans, Reformed, Presbyterians, Methodists, Baptists, Pentecostals, and so on. To the degree that these "specialisms" express a world the riches of which cannot be embraced by one person or group of persons, so that a multiplicity of ways of confessing and believing becomes possible, this is even to be seen as a positive factor. But that is to put things very idealistically. The way in which the denominations came into being in the past and absolutized their own "truth" has added many shameful pages to church history. Fortunately, in our time the divisions are constantly being relativized, as a result of which the one Denominator is again becoming visible under which all the variants can be brought, so that, as algebra teaches us, the amounts in the numerator can be added up to a positive sum.

Another distinction also needs to be made in the denominations, as in physics. There are doers and thinkers, practical Christians and mystics, orthodox and progressive people. There are Christians who live in the world with wife and children, with their work and material cares, and who in all this try to lead the life of faith they form by far the great majority. And there are those - though they form a gradually dwindling minority - who voluntarily renounce all that ties them to earth and seek God through the mystical way of repentance, prayer, meditation, and choral offices.

If it works out well, each supplements the other. That is not as impossible as it might seem. Thomas à Kempis, who was a religious, wrote his book six hundred years ago specially for the discipleship of his fellow monks. It is still read today, also and above all by those who live in the world, because it contains hints which help them. Modern monasteries open their doors to people who, weary of their worldly way, want to withdraw to get their breath back and gather new courage.

Finally, it is quite exceptional for the two types to be found united in a pure form in one person. Blaise Pascal was such a person: Apart from being a brilliant theoretical scientist and experimenter, he was also a mystic who until his dying day worked for the poor of Paris. Anyone who thinks that that sort of thing can no longer happen in our day would do well to spend some time considering the life of Simone Weil (1909-1943), a young French woman who has much in common with Pascal.

I said that the way of the physicist is exciting and fascinating. Those words aren't enough to describe briefly the emotions on the way of faith. That is something which I can't do. I think it is because these emotions are aroused at a deeper level of existence. To get some idea of this, perhaps the best thing is to work through the book of Psalms, the only collection of poetry which the Bible is rich in, from beginning to end. Everything can be found there. Tranquillity, trust, surrender: "The Lord is my shepherd, there is nothing I shall want" (Ps. 23). And also rebellion: "I will say to God, Why have you forgotten me?" (Ps. 42). Longing for the experience of God's presence: "Like as the hart desires the water-brooks, so longs my soul after you, O God" (Ps. 42). The peace of the one for whom this has become reality: "You have fashioned me behind and before: and laid your hand on me" (Ps. 139). But also the complete opposite: "My God, my God, why have you forsaken me" (Ps. 22). Joy at the way which God shows through his commandments: "Your word is a lantern unto my feet" (Ps. 119); but also bitterness about the prosperity of those who snap their fingers at this way (Ps. 73). Unadulterated hymns of praise, too many to name, at the wonder of creation and the place of human beings in it: "What is man that you are mindful of him?" (Ps. 8). But also dismay at one's failings: "Be gracious to me, O God" (Ps. 51); and despair at a deadly sickness: "I am at my wits' end" (Ps. 88).

This is only a selection, but it also makes it clear that one description does not fit this way: pleasant, attractive. Why then should one choose it? This question is very difficult to answer, and there are just two things about it that I want to say. First, that we are literally grasped by it; we cannot get away from it. And secondly, that this way opens up a prospect on a future which is called "the kingdom of God": a new heaven and a new earth in which righteousness dwells. That kingdom does not come through our efforts (that is where it differs from the ideologies whose failure we are seeing in our time), but it does not come without us either. We may be fellow workers, no more but also no less, in a work which is worth the trouble and of which the long-term fulfillment is sure.

What do I believe? A long list of truths, dogmas, miraculous facts? I believe - and I write this with the greatest possible hesitation and modesty - that I am on the good way. And because you can wander off it before you know that you have done, no prayer is more relevant than the closing lines of Ps. 139, which I have written at the head of this chapter.

11 Intersections

"I am a sojourner on earth; hide not your commandments from me!" Ps. 119:19

Collisions, confrontations, often take place at the intersections where the ways of science and faith meet.
Sometimes - though that is now long ago in the past - the
church interfered with science, as in the case of Galileo. In
the more recent past and in the present, believers have been
put on the defensive. They have to defend themselves
against arrows shot against them by science (or what passes
for science). In an earlier chapter, Chapter 7, we came upon
instances of this, questions like "Is biblical belief in creation
compatible with the results of physics?" In this chapter, I
want to discuss two other questions which are connected
with the relationship between faith and physics. The first
is, how do physicists who are believers practice their discipline in relationship to their faith? The second is: Has faith
anything to do with the social consequences of science and
technology?

MIRACLES

One of the most scandalous things a scientist can do is to
believe in miracles. "Don't tell me that you swallow all that
hocus pocus," you sometimes hear people say. What they
mean is someone who walks on water, feeds five thousand
people from five loaves and two fishes, and still has some
over, and who changes water into wine (turns H_2O into
C_2H_5OH).

What causes offence about these miracles is that they are
in conflict with the laws of nature (the law of gravity, the
law of the conservation of mass). In general, miracles are
events which a person had thought not to be possible. It
follows from this that the question whether a particular

event is to be regarded as a miracle or not gets purely subjective answers. It depends on what the person in question thinks possible. Suppose that someone who died before 1900 rose from the grave and looked round our world, he or she would see numerous miracles: "the miracles of technology," like the telephone, the radio, television, aircraft and spacecraft. But, Paul Davies explains, these are not real miracles, since they all rest on the laws of nature and thus can be explained rationally. That is not the case with miracles that may not be.

Why may they not be? In the discussion in Davies' book between the "skeptic" and the "believer," the believer comes off badly because he operates at such a deeply simplistic level; but the skeptic, too, ultimately offers no proofs. Davies comes to the rather feeble conclusion that the scientist prefers not to believe in miracles, because "the scientist [...] prefers to think of the world as operating to natural laws." That can hardly be called an argument; it is a matter of personal preference connected with the ideology of the scientist, a question of faith, if you like.

In Christian circles, the attacks of science on belief in miracles, which are becoming gradually less serious, are usually parried like this. First of all, it is observed that science is exclusively concerned with repeatable phenomena, and that the laws of nature are formulated on this basis. Unique events, by definition, fall outside this. Jesus is no longer available to repeat his "experiments" before the critical eyes of scientists, rather than those of credulous disciples.

Secondly: Why should the laws of nature be sacred? Doesn't the God who made them have the power to suspend them at times? Of course, he must not make a habit of this because, as we saw earlier (in Berkhof's view), the reality around us is no "haunted house," precisely because of its regularity, but a trustworthy framework created by God in which we can move. But, Herman Berkhof says, if it may not be a haunted house, it is not a bunker either. There must be room in it for miracle.

We ought to be able to stop there, because any attack by

science on faith has already been disarmed. But I shall go on, because this result does not satisfy me: First of all, because it is again so defensive, and secondly, because in this sphere miracles again take on too much the character of "truths," facts that are held to be true, something that is not forbidden on scientific grounds but beyond which one cannot go much further.

In the Bible, miracles are almost always called "signs." Signs indicate a *sign-ificance*. To the degree to which this significance gets through to us, the miracles become true for us. This significance is usually not that the laws of nature are violated, but that something happens that we had thought impossible in human terms. I shall try to illustrate that from the story of the marriage at Cana, where Jesus turned water into wine.

At that feast, where Jesus and his disciples were guests, the wine ran out half-way through. The feasters and the master of ceremonies panicked. How can a feast go on if there is no more wine? Then Jesus intervened. He had six stone water pots filled to the brim with water, and then asked the master of ceremonies (who did not know what had happened) to test them. In amazement the latter asked the bridegroom why he had kept the best wine until last. "And his disciples believed in him," the evangelist John adds in conclusion (the story does not appear in any of the other Gospels).

What is the significance of this story? First, that Jesus sets a powerful example. Here is a demonstration of power, by which he shows how he can control even the iron laws of nature. The bystanders are perplexed. Someone who can do this sort of thing must come from God. Jesus overwhelms them with a demonstration of supernatural power.

That is what I do not like at all about this interpretation, because it is in conflict with the picture that the Gospels paint of him: He is not concerned to show his power but to show service. So I think that the story is seeking to convey something else to us, namely that a feast which is kept going by human means like wine will be interrupted if the means are exhausted. That feast is an image of human existence, which ends up in a hangover if the human stimulants prove inadequate. It is a feast at which the wine runs out half-way through, and sometimes even earlier.

That feast, the story says, can take place at a higher level, with a deeper joy (the new wine was better than the old), if the guests drink from the pots which were filled to the brim by Jesus with water. "Living water" is the image that Jesus uses for what he has to offer, also right at the beginning of the Gospel of John: his words, himself. "Whoever drinks of the water that I shall give him will never thirst; the water that I shall give him will become in him a spring of water welling up to eternal life." That is what the wedding guests have experienced as a miracle because they had not thought it possible: a preliminary proof, a sign of the coming kingdom in which God will be all and in all.

That is only a personal view, a piece of amateur theology if you like. So I shall go even further. For me, the story could also end like this: "Among the guests there was also a chemist. He was going about there in his white jacket, and when he saw the throng round the water pots his curiosity was aroused. He quickly took a sample and took another one after the so-called miracle. 'Just as I would have expected,' he mumbled, 'H_2O both times.' He was the only one of the wedding guests not to notice the miracle."

I hope that you won't think that this explanation is a clever attempt to talk the offence out of the story. I don't feel the least need to do that, because as far as I am concerned this offence (the violation of the sacred laws of nature) does not exist. But the story takes on content for me only in the way which I have described. I shall introduce a second brief illustration to show that for me it is not a matter of demythologizing the miracles as an end in itself.

The resurrection of Christ is the central miracle in the Christian faith. Both friend and enemy are, I think, agreed that Christianity stands or falls by it. Recently, discussion has again erupted within the Christian churches as to whether the resurrection must nowadays be understood in literal, physical terms ("the empty tomb") or spiritually: Jesus lives on in the spirit of his disciples, who express that miracle in their Easter stories. One can also stand above the discussion, as Malcolm Muggeridge does:

> I am sure there was a Resurrection, but I don't in the least care whether the stone was moved or not moved, or what anybody saw, or anything like that. I am absolutely indifferent to that. *But there must have been a Resurrection because Christ is alive now.* He is alive now in the sense that he exists now as a person who can be reached. His life is still valid, so that it is possible, not only to hear and learn, but *experience*, the truths that he propounded.

For me, that puts into words the essence of the resurrection event. But I would add that the account of events on the first Easter morning as given in the Gospels does seem to me to be credible, particularly because of the utter unbelief in the reactions of all those who had not yet had their noses rubbed into the facts. They had to be convinced bit by bit, with the greatest possible reluctance. These stories seem to me so authentic that I don't feel any need to spiritualize them. What I find a weak point in the writings of theologians who think they still have to do that is that they do not make it clear why it is so necessary. Or do they, perhaps unconsciously, still think that the story has been so undermined by science that it is no longer fashionable to accept it? I would very much like to relieve them of this worry.

SCIENCE AS A CULTURAL FORM

Here are two more lines of the poem by Jacques Bloem from which I quoted the first two earlier. The first verse goes:

> Is it enough, one or two poems,
> to justify an existence
> spent badly performing foolish duties
> gradually wasted for all too scanty bread?

Jacques Bloem was a Dutch lawyer, and in everyday life he had very modest duties in legal administration. He did his work without a trace of ambition or enjoyment, and with great reluctance: foolish duties to earn a sparse living, wasted time. Outside that, he lived for what he was born to, poetry.

There are thousands of people who, like Bloem, spend their days in uninspired work which makes no appeal to their creative potential. It always makes me somewhat ashamed that this has been spared me, and that in addition I am also paid, and not a sparse amount. In this book I have tried to give some indication of the joy that the explorations of physicists bring them now and then.

I must say one thing at the start. There is no such thing as Christian physics. The rules of the subject are the same for everyone who practices it. We do it together, and most of the time we do not know whether our colleagues are believers or not - nor is that a matter of prime importance. Only when you reflect on what you are engaged in, when you try to fit it harmoniously into the rest of your life, does belief play a role.

As a physicist, I am occupied in a world created by God: in his garden, as we read at the beginning of Genesis. Human beings, too, are created by God and are created - we are explicitly told - *in his image*. Now given that the first thing we are told about God is that he is creator, being "in his image" also means that human beings are meant to be creators. That means that the creation as it emerged from God's hands, and which God found to be very good, was not finished, and perhaps that is also why it was very good. Human beings can only show the image of God if there is something left over to create after his creation. I like to compare that with toys. Many of the modern toys with which we can still surprise our children are finished. You can get just one preprogrammed thing, and after a day there is no more enjoyment in it. There is nothing more to experience. It's a worthless creation. But a box of Lego is a creation with which children can do something, an endless source of possibilities. That is also what God's creation looks like.

The notes of music were in it, the letters of the alphabet, the colors of the palette. The symphonies, the poems, and the paintings that can be made from it are inexhaustible. All the statues that were ever sculptured sat hidden in formless lumps of stone and were called forth from them. The wheel was there and the steam engine; Newton's laws and the theory of relativity, hidden in apparently unordered matter. Are they also creations? Certainly. Creation is a matter of calling into being what was hidden.

In the Genesis story, we read that God rested on the seventh day from the work that he had done. That means that he must have been wearied by his exertions. That seems surprising, because in the preceding narrative we simply have, "And God said," and it was there, apparently without difficulty. But that was evidently not the case. God was weary, exhausted. From looking for the right word? I recognize the description of this process of creation, including the joy on the "seventh day." That's good. That's far more than self-satisfaction. We are evidently included in this joy as creators in the image of God.

Physics, poetry, sculpture, technology, and so much more art works of *culture*. "And God put man in the garden to till it and keep it." Cultivating the garden is culture. Making it blossom out of stony ground. Cultivation is very different from exploitation: getting out what is in it, regardless, despite the consequences for the garden. We come up against this truth in modern times, when our world threatens to break down under the pressure of our exploitation. That is cultivation without bringing out new possibilities, cherishing, looking after, tilling and keeping. Has Genesis really been overtaken by science? In our time it seems, rather, that the way in which we do science and technology has been overtaken by Genesis.

Culture brings forth works of great beauty: our vineyards and orchards in which every tree seems to stand in its place. Poetry in which there is not one word too many. The almost empty paintings of Mondriaan. Maxwell's laws, pearls of mathematical beauty and expressive power. Simplicity, wrested from a chaos of phenomena, in which Newton recognized the simplicity of his creator. Miracles which reduce a person to marvelling.

Finally, culture and cult belong together. The significance of the Latin word in which both are rooted is "till, care for, revere." The human beings who till and care for the garden, at God's command and in accordance with his precepts, honor the Creator with their work. This honor can also be offered to him in a spontaneous song of praise. The poets of the Psalms did it for us earlier: "How great are your works, O Lord, you have done them all with wisdom."

I can easily imagine people who read what I have just written shrugging their shoulders and saying, "Why do we need all that? Remove God from the story, and what we are left with is the fact that creative people can find great joy in their creative work, be amazed at the beauty of culture. I have that experience myself. I don't need God for it. And we can also see for ourselves how we are polluting the environment. We must do something about that."

I can't refute such people or prove that they are wrong. Nor am I concerned to show that I am right and others wrong. What I want to make clear is that it is necessary and meaningful for me to look at the practice of physics in this way, because this activity has to have a place in the whole of my life. That life must be an answer to God's question: "What are you doing, and why?" I am responsible: I have to give a response. Now of the fifteen hours a day at my disposal, I devote at least eight - according to some people too few, but perhaps far too many - to physics. These cannot and may not be "detached from God." I know that that isn't simple. The lofty prose that I wrote above is not clear to me every day. In addition, the practice of science can have other aspects for human beings and society which can evoke doubt and disquiet.

CONSEQUENCES FOR HUMAN BEINGS AND SOCIETY

Recently I had a visit from a student who came to ask me to take part in a conference of her student union on the subject "Ethical Problems of the Physicist," or something like that. My task was to be to discuss one or more cases from my research where I had had to say, "I cannot take part in that research for ethical reasons, because of my Christian convictions."

I had to disappoint her, though in principle I like to accept this kind of invitation. The reason was simple: I had never found myself confronted with such a dilemma in my thirty-year career of research. I hope that that has not been because I am a scientific idiot who goes around with ethical blinders. My experience coincides completely with what the physicist Casimir describes in the last chapter of the book I have quoted from earlier as the "science-technology spiral." It should be required reading for all students in science. I shall give a summary of the conclusions, leaving out the well-documented arguments which Casimir gives in support of them.

The results of pure, basic physics find their application in technology. However, that applies only to a tiny fraction of these results. Moreover, technology never uses the most recent, and rarely the most profound, results of academic research. Technology lags around twenty years behind. That means that academic researchers do not and cannot have any idea whether their results will ever lead to technical applications, far less what these might be, in such a way that they could be worried about the social applications of their work. Conversely, the latest novelties of technology are used in scientific research almost immediately, without any lag. Only the laboratories which are in a position to get quickly the newest computers, the most recent electron microscopes, and so on, can keep ahead in the race.

The most important feature of this scientific technological spiral is, as Casimir indicates, the fact that it is an autonomous mechanism. It works in precisely the same way in countries with totally different social systems, like the liberal system and the Marxist system. There is only one answer to the question "Who controls this spiral?" - and that is, "No one." The progress of the spiral can be accelerated or slowed down by government regulations, wars, or revolutions, but these are always local and temporary by nature. The basic progress is not influenced by them.

It is therefore hard to make the practitioners of fundamental physics responsible for the consequence of their work. I would point out in passing that at this point researchers behave somewhat inconsistently. If in their research proposals they appeal to the pots of money which the sponsors have, they do not neglect to hold out the prospect of mountains of gold in the form of possible applications, mentioned by name, in the not too distant future, which acceptance of their proposals will produce. Ad Lagendijk, a professor of experimental physics at the University of Amsterdam, who in his inaugural lecture called this sheer deception (he even used the word "prostitution"), with a reference to Casimir's spiral, did not increase his popularity among his professional colleagues in so doing.

But, you might perhaps say, what I have just written may be true of the pure, fundamental research which does not have any direct technical applications, but things are different for the research which is carried on in industrial laboratories and at technical universities. Surely people there have clear, concrete applications in view, which can be realized in the short term, and one can judge whether these serve a good or a bad end? If only it were so simple! Here is an example.

In the 1940s, building on earlier basic research into the semiconductors germanium and silicium, in industrial laboratories the transistor was developed. This transistor has started a revolution in the electronics industry in our century. It has given the world a different face. It is no exaggeration to say that after the stone age, bronze age, and iron age, we have now entered the silicon age. Among other things, modern computers emerged from this. Could the inventors of the transistor in any way see what the consequences of their work would be? Not at all. We cannot even assess it today, after the event. Is the computer a blessing, or a curse, or both? Who knows? My own answer is that it is both, but I do not have a balance on which I can weigh up both the positive and the negative consequences. The transistor, the television, the computer are products of the spiral; they were unavoidable, and they cannot be done away with. In themselves they are neither good nor bad. Their makers need not reproach themselves; at most, they should perhaps be somewhat less proud of their work.

But - and here is the last objection - surely there is one area where matters are clear: the armaments industry. Down the centuries (already beginning with Archimedes), physicists have been heavily involved in making weapons. The atomic bomb was based on their own discovery of the process of splitting the atom, and they themselves produced it. Surely that - and much else - should never have happened? Sadly, things are not so simple either. Only pacifists have no problem here, and pacifists form a very small minority. Most people think that organizing armed defence against possible attacks from outside is permissible and necessary. Most Christians - myself included - hold precisely the same view. So for them, arms manufacturing is not in principle a dirty job. That it is, nevertheless, thought good form for many people to speak with contempt about arms manufacturing (and also about professional soldiers who organize armed defence in our name) is just a form of hypocrisy.

As an example, here is a brief account of the origin of the atomic bomb. A uranium nucleus was first split successfully in 1938 by Otto Hahn and his colleagues, thus releasing energy. From that moment, in principle there was the possibility of giving this process explosive character through a chain reaction. The only problems which had to be overcome were of a technical kind: gigantic, but not insuperable. In 1941, through Niels Bohr in Copenhagen, the Americans received the deeply disturbing report that the top physicists left in Nazi Germany, including Heisenberg, were busy constructing a nuclear bomb. The choice was thus either to wait for that to happen, in which case Hitler could be assured of world rule for a long time, or try to get there first. For Einstein, more than most a man of peace, the choice was not difficult. He threw his full weight into the scales in order to persuade President Roosevelt to set the process of production going. This led to the greatest organized effort of physicists, chemists, and technicians in history. The outcome is well known. I have studied this history thoroughly enough to dare to say that if at that time, in those circumstances, they had asked me to join in, I would have done. I would still support the decision today. Unfortunately, choices in the world are not always between good and evil, but often enough between evil and greater evil.

What I have said does not mean that there can never be a moment for scientists when they have to say, "I cannot join in this." Anyone who collaborates in something as dirty as the development of a fragmentation bomb is a criminal; there are no words to describe the makers of the explosive toys that the former Soviet Union dropped on Afghanistan to blow the limbs off children. But these are striking exceptions. If scientists have become involved, can they not at least do everything possible to limit the damaging consequences?

Examples can be given in which that has been attempted, but they are sparse, and the effect was virtually nil. Directly after the making of the first atomic bombs - by then, Hitler was already dead - a large group of those involved in the project attempted to prevent the immediate dropping of the bomb on Japan. Their argument was, "First enlighten the Japanese; if you like, demonstrate the devastating power with which they will now be confronted if they don't capitulate."

No one listened to them (I leave open the question whether they should have done). Directly after the war, there were earnest and passionate warnings by Bohr, Einstein, and others against the possibly apocalyptic consequences of the arms race which had begun, from the hydrogen bomb to star wars. Nothing came of them. Can one without further ado nevertheless condemn all those scientists and technicians who took part in these programs? That seems far too easy to me. I am not their judge, but if I had to be, then it would have to be in a fair trial, in which there were prosecution and defence; in which all the facts were put on the table, and all the motives tested in the context in which they arose. It is quite possible that then at least we could come to some understanding.

What conclusion are we to draw from all this? The work of scientists has unmistakable social consequences, positive and negative. In by far the great majority of instances, these consequences are outside their reach. They cannot be foreseen, and if they could be foreseen, they could not be weighed. Once they happen, they cannot be undone. Must we end our survey with this powerless conclusion? What is this account really doing under the heading of "intersections," the intention of which was to investigate where the ways of physics and faith intersect?

At the end of the chapter on the spiral of science and technology, Casimir sees himself faced with the same block. "I could end my book here," he says, "but I feel obliged to go further. In my young years, my colleagues and I took the possible social consequences of our work lightly. I don't want to make this mistake again in my old age. I want to express something of my doubts, my fears." In what follows he bears witness to his deep disquiet at the pollution of the environment and, above all, at the terrifying accumulation of armaments. This is much more a cry from the heart than a recipe for how the evils are to be cured. His answer to the question "What can we do as *physicists?*" is "Virtually nothing. We can only try to behave as adult citizens, and indeed as well-informed citizens, not only as citizens of our own land but also as world citizens: Our many international contacts can contribute to that. We can help to challenge erroneous ideas and we can try to expose hypocrisy and misunderstanding. But we alone cannot either stop or control the disastrous spiral." It is a poor answer, and he knows it.

From these honest pages, the integrity of which shines out brightly, I want to quote three passages which put me on the track of pursuing this view further. Here they are:

> Traditional religions contain an explanation of the origin of our world and of life on earth; they also contain an ethical code. Darwin's *Origin of Species* and Wein-

berg's *First Three Minutes* have driven out Genesis. This repudiation of the biblical images of creation has led many people also to begin to doubt Christian moral principles, and there science can offer no help.

We have made science our God.

Science and technology form an admirable and productive pair, but they have detached themselves from all bonds and limitations, including those laid upon them by wisdom and love of neighbor.

These three passages sum up in brief what the age of the Enlightenment has achieved. Physics has rejected the traditional picture of the world, the cosmogony, of the religions. From this, countless people, inside and outside science, have drawn the conclusion that none of the other things that the religions have to say need be taken seriously any longer. Religion can be written off. Human beings have become autonomous. From now on, they have to choose for themselves, and science is their guideline. Science will have to provide the foundation for a new world view and show people the way of Progress. There are no longer any intersections between the ways of belief and scholarship: The last intersection lies two centuries behind us, when the ways definitively flowed together into one, that of science. That is a social consequence of the first order: Science has detached human beings and their world from God and even has taken God's place.

It is at this point that physicists who also try to be believers in our time can and must do something as *physicists*. They must make it clear that this way is a dead end, that the expectations aroused are *false*, that they rest on sheer megalomania, on arrogance. Believers must not belittle science, but show science its due place, that of a *servant* of humanity and not its idol. That is what I have tried to do in this book.

What I have said so far leads to another conclusion, namely this: Perhaps the spiral of science and technology is not so autonomous, and the outcome is not as unavoidable as Casimir suggested. One of his main arguments is that the spiral works in precisely the same way in two totally different social systems like the liberal system and the Marxist system. That argument loses much of its validity when it can be demonstrated that these two systems are not as different *in their aims* as might seem at first sight. (Vaclav Havel puts it like this: "The Marxist system reflects the capitalist system in a distorting mirror.")

Perhaps the choice is not limited to the two systems *as they now function*. And even if we have to conclude - and in the 1990s, who does not - that the Western economic system (a free market with a social safety net) is superior, that does not mean that it must continue to exist throughout the whole world as it now does in the West.

In the Western system, it seems, the outcome of the spiral, the production of goods, is determined by what the majority of people want. The questions which then arise are, for example: What do we really want, and why? Is what we really want happening? Or can and must things be different? That is what the conclusion of this book will be about.

PERSPECTIVE

Someone has thought up the term "the me age" for the climate of the 1980s, which are just behind us. This was a time in which the main concern of many people seemed to be how they could enjoy themselves as much as possible. Nowhere have I seen this mentality summed up more concisely than in a TV commercial which is a model for all others. A woman who wants to sell us a particular brand of drink sings a song which ends with the line "I do what I like." Unless the signs are deceptive, that period did not end on I January 1990. What are the prospects, the challenges for the 1990s? Here are three which I have picked out of recent newspapers.

The problem for the food industry is that the consumer

has a limited capacity. Full is full, and even advertising campaigns vainly press against this limit. So a successful attempt has been made to increase the profit, not by increasing the turnover in tons of food but by disposing of the same number of tons in luxury packaging. There is also a limit to that, of course, but now, I read, the year 1990 promises to see a breakthrough. Artificial fat is coming, along with a number of other "foodstuffs" the property of which is that they do not feed and yet are attractive, and fortunately are also expensive. The consumer can eat and eat, and volume and profit can again be very high. The ancient Romans used to put their fingers down their throats when they were full, but thanks to the combined efforts of science and technology, this tasteless trick is no longer necessary. When now and then - preferably not too often - we see the hungry children from Africa on our screens, we can always give a coin. Hopefully, in the future, commercial television will spare us such distasteful pictures. Unless, of course, market researchers show that seeing such people now and then is part of our pattern of needs.

Another great challenge for the 1990s is, I read, the development of a television picture with twice as many lines as those on the present sets. That will require an enormous effort, but the result will be the quality of the cinema in a domestic room, also combined with good stereophonic sound. We needn't worry about the quality of the programs; that is guaranteed by the owners of the commercial television stations, who know what is good for us and certainly aren't concerned to give us unpleasant moments. Of course, the rooms in our homes are already all too full. The stereo speakers and VCR are already there. The telephone with a screen is on its way. In addition there is the personal computer, which before long will also rescue us from bothersome shopping. We shall then call up various stores on our screens, tick our requirements, and these will be sent to us quite automatically. Then we shall sit down, each in our own cell, with everything within reach, in a room full of means of communication. And sometimes there will still be complaints that many people in our society are so lonely.

Perhaps that is also something for an advertising campaign: "Ring up your neighbors. See their face on your screen."

Are there then no "higher things" in prospect? Certainly. On taking office, at the beginning of 1989, President George Bush formulated a challenge that was again to give humankind a longer-term goal: a manned space flight to Mars. There was a provisional, cautious, price tag on it, 200 billion dollars: $200,000,000,000. That's a lot of money to spend on activity in space, certainly when compared to what can be done with it on earth.

Is that what we want? At first sight, it certainly seems so. The Western, liberal democratic system works on the principle of the free market. Market research sounds out what people want, scientists and technologists make it possible, manufacturers produce it, and the customer buys it. The circle is closed, the desires of the consumer are satisfied. The failure of the Marxist economy that we are seeing nowadays rests on disregard for this cycle. When the Wall came down, the first place to which thousands streamed over from the other side was the Kurfürstendamm, where the abundance (why not say the debauched luxury) of the West is displayed in the shop windows. That is evidently what they also want. So do a billion Chinese. The triumph of Western ideology is total. There are already people in the West who are worried that from now on life will become rather boring. There is nothing more to fight for or to prove; the superiority of their own system has been established, and virtually no one doubts it any more. The only thing left is to live out all one's years and to consume a few per cent more each year.

Liberalism and Marxism may perhaps seem to us to be two completely opposite poles, but they are less different than one might think. They derive from the same root: nineteenth-century belief in the Progress of the autonomous human being who, apart from God and his commandments, has taken the way to paradise on earth. "This paradise knows no forbidden fruits," rejoices an advertisement which wants to sell me some attraction. The way of Progress in both systems was pioneered by science, elevated to a god. Marxism-Leninism is itself purely scientific, we are assured, and anyone who serves other gods is an opium pusher. Nowhere in history has belief in science been so unconditional as in that system.

But Western economic liberalism also has its starting point in science: the philosophy of Adam Smith from the eighteenth century, in which the market mechanism is central. This promises freedom and unlimited growth, made possible by the link between scientists and technologists. If we ignore a few problems, like economic crises, millions of traffic victims, and the state of the environment, this promise has largely been kept.

Is that what we want? What do we really want, and why? Do we, for example, so much want a man on Mars, to provide us with the excitement that we desperately need in our attractive but boring existence as consumers? I find that hard to believe. Suppose that you were to give a leaflet to a thousand random passers-by and asked them to write down a couple of things that they would like for the years to come, for themselves and the world? How many of them would say "A man on Mars"? Of course I don't know, but I think that the number would be negligible. I think that we would get quite different replies. Such as: I'd like to be healthy. I'd like to be a bit less lonely. I'd like to give and receive more love. I'd like to know better what to do with my life. I'd like peace, an end to injustice. That sort of thing - I hope. For I hope that despite all the manipulation to which our souls are exposed, we have not lost the vision of another world, of a truly human existence.

For we are manipulated, extremely subtly and often imperceptibly. By politicians, for instance, who try to persuade us to spend our tax money on their large-scale space projects. By the advertising industry, which tries to make us want and buy industrial consumer goods. Some people even say that we don't have these needs at all, but that they are implanted in us by advertising and subsequently satisfied by our purchases. I don't believe that that is true. I think that things go deeper, that we do really have this desire for comfort and pleasure. Advertising does not create it but appeals to it. It brings it out, so that it begins to dominate us and to throttle the other feelings that we have.

What do we really want? The dilemma has been expressed acutely by George Steiner, the linguistic philosopher. In the interview I quoted earlier, he says:

> Television brings a greater mass of information about injustice, terror, hunger and sickness into every home than ever before. Not to know is now a positive act. You can't say "I haven't heard." What you can say is, "There's nothing I can do about it." Yes and no. If a Gandhi or a Mother Teresa had said, "There's nothing I can do," then all kinds of historical changes would not have taken place. But why should we look only at the "giants"? In every city in which people like us live, there is an urgent need for people to care for the sick and the aged: not great doctors or therapists, but someone to watch by a terminally ill patient or to take an old person to the WC. No special professional knowledge is required for that, but enormous dedication of heart and soul. So you might say, "I can do nothing about El Salvador, Cambodia or the Gulag Archipelago," but often round the corner you can begin something really important. On a modest scale you can do something to alleviate this daily misery and loneliness.

And when the interviewer asked, "And what if we don't make this choice," Steiner replied:

> If you don't, that is a deliberate choice. You decide that your own comfort, taste, reward, career, enjoyment, are the most important things for you. Because millions of decent people like us make that choice every day and forget that we are making it, you can perhaps say that the process of dehumanization is going on at a large-scale level.

Decent people like us, says Steiner. He includes himself, and I echo his words. Which of us escapes this? But also, in which of us, though perhaps only weakly, do such words not find an echo? I think, or perhaps I just hope, that something is alive in every person of the sense that we are not created for a pleasant life in which all turns on comfort, reward, career, enjoyment. That we cannot be at home in a world with these concerns. For that is what was meant by the author of the psalm the words of which I put at the head of this chapter, "I am a sojourner on earth." And what he adds is not a warning with uplifted finger which tells us what must be. It is a *prayer* of someone who knows what direction to look in, and also that he is not doing this in his own power. This prayer is, "Hide not your commands from me."

Bibliography

Andreus, Hans: *Gedichten (Poems) 1948-1975*. Haarlem 1975

Andriesse, Cees D.: *De Diefstal van Prometheus (The Theft of Prometheus)*. Contact, Amsterdam 1985
-, *Een Boudoir op Terschelling (A Boudoir on Terschelling)*. Contact, Amsterdam 1987

Berkhof, Herman: *Christian Faith* (1973). ET Eerdmans 1986

Bohm, David: *Wholeness and the Implicate Order*. Routledge and Kegan Paul 1980

Capra, F.: *The Tao of Physics*. Shambala, Berkeley 1975

Casimir, H.B.G.: *Haphazard Reality - Half a Century of Science*. Harper and Row 1983

Davies, Paul: *Superforce*. Simon and Schuster 1984
-, *God and the New Physics*. Penguin Books 1983

Einstein, Albert: 'Science and Religion'. *Nature* 146, 1940, 605
-, B.Podolsky and N.Rosen, 'Can Quantum-mechanical Description of Reality be Considered Complete ?'. *Physical Review* 47, 1935, 777.

Gerhardt, Ida G.M.: *Verzamelde Gedichten (Collected Poems)*. Polak and van Gennep, Amsterdam 1985

Gleick, J.: *Chaos*. Penguin Books 1987

Green, Julien: *Frère François*. Editions du Seuil, Paris 1983
- , *De Aansprekers (The Undertakers' Men)*. Arbeiderspers, Amsterdam 1979

Hart, Maarten 't : *Een vlucht regenwulpen (A Flight of Curlews)*. Arbeiderspers, Amsterdam 1978

Hawking, Stephen: *Is the End in Sight for Theoretical Physics?*. Cambridge University Press 1980
-, *A Brief History of Time*. Bantam Books 1988

Hermans, Willem Frederik: 'De kleurentheoloog' ('The Colour Theologian'). *NRC-Handelsblad*, 25 March 1988

Hilgevoord, Jan: 'Holisme in de natuurkunde'. *Nederlands Tijdschrift voor Natuurkunde 53*, 1987, 128

Jammer, Max: 'After Dinner Address'. *Philosophical Magazine B 56*, 1987, 1055

Kolakowski, Leszek: Interview in *De Tijd*, 10 February 1989

Lagendijk, Ad: *De arrogantie van de fysicus*. Inaugural Lecture, Amsterdam 1989

Manuel, Frank E.: *The Religion of Isaac Newton*. Clarendon Press 1974

Marquez, Gabriel Garcia: interviewed in English by the Dutch TV station, VPRO. The interviews were translated and published as *Nauwgezet en Wanhopig*, VPRO 1989

Muggeridge, Malcolm: *Jesus Rediscovered*. Fount Books 1969

Mulisch, Harry: 'Het licht'('The Light'). *NRC-Handelsblad*, 12 February 1988

Nolthenius, Helène: *De Man uit het dal van Spoleto*. Querido, Amsterdam 1989

Pais, Abraham: *Subtle is the Lord: The Life and Science of Albert Einstein*. Oxford University Press 1982
-, Interview in *Trouw*, 5 April 1989

Pascal, Blaise: *Pensées*. Everyman edition, Dent 1908

Steiner, George: in *Nauwgezet en Wanhopig*, VPRO 1989 (see above)

Thomas à Kempis, *The Imitation of Christ*. Penguin Books 1952

Van den Berg, Jan Hendrik: *Gedane zaken (Things Done)*. Callenbach, Nijkerk 1977

Van der Hoeven, P.: *Newton*. Het Wereldvenster, Baarn 1979
-, *Blaise Pascal*. Het wereldvenster, Baarn 1964

Van der Meer, Simon: Interview in *NRC-Handelsblad*, 18 April 1987
-, Interview in *Hervormd Nederland*, 25 February 1989

Van der Zee, William: *Ape or Adam?* Genesis Publ.Company, North Andover MA 1995

Verschuuren, Gerard: *Life Scientists*. Genesis Publ.Company, North Andover MA 1995

Weinberg, Steven: *The First Three Minutes of the Universe*. Andre Deutsch 1978

Zukav, Gary: *The Dancing Wu-Li Masters*. William Morrow, New York 1979

Index

OF RELATED INTEREST

Ape or Adam?
Our Roots according to the Book of Genesis
by
William Van der Zee

This book is the result of a series of radio lectures on the book of Genesis (1-12). Creation is not a matter of the past. We are right in the middle of this process. At this very moment, God is being involved in creating man to His image and likeness. At this very moment, God is being involved in making a world in which it is good to live. And the seventh day is ahead of us, the great day of the Lord. Tomorrow it will be sabbath, tomorrow I will be free.

If the bible is not a final authority in matters of science and doesn't offer a theory as to the genesis of the earth, of life, and of humankind, it is impossible for the book of Genesis to come in conflict with any scientific theory whatsoever - just like in turn no scientific discovery is able to dethrone God or to refute the biblical testimony. You come to the sudden awareness that bible and science, faith and knowledge do not contradict each other. Even the concepts of creation and evolution fail to come out as absolute contrasts.

6" x 9"; 120 pages with index
ISBN 1-886670-03-X softbound $ 19.50

Genesis Books are available at special quantity discounts for fund raising, educational use, sales promotions, or premiums.For detail write, fax, or telephone the Department for Special Markets, Genesis Publishing Company, Inc., 1547 Great Pond Road, North Andover, MA 01845. Tel. 508 688-6688; Fax 508 688-8686.

OF RELATED INTEREST

Life Scientists

Their Convictions, Their Activities, and Their Values.

by

Gerard Verschuuren

Life scientists seem to be the wizards of the new age. They dared to tackle issues ranging from the origin of life to the origins of humankind; from medical and genetic knowledge to questions about how to control life and death. Spectacular are the achievements they have made. Who are these scientists? How do they operate? What do they believe? What do they value?

Science is no more objective and rational than the humans who do it. This book shows us the 'world' of life scientists, seen through the eyes of a biologist with a philosophical background. It's a guided tour through the philosophy of the life sciences, based on the insights of modern scholarship, without going into unnecessary technicalities. Science on display, so to speak, with the intention of making the science world more transparent. In reading this book, you may learn a little *of* the life sciences and a lot *about* the life sciences.

6" x 9"; 288 pages with index
ISBN 1-886670-00-5 hardbound $ 34.50

Genesis Books are available at special quantity discounts for fund raising, educational use, sales promotions, or premiums. For detail write, fax, or telephone the Department for Special Markets, Genesis Publishing Company, Inc., 1547 Great Pond Road, North Andover, MA 01845. Tel. 508 688-6688; Fax 508 688-8686.

"This book [...] demonstrates the author's extensive and profound knowledge of biology. I enjoy the (brief) historical context in which he places the issues at hand and the clear-headed philosophical analysis. The writing is succinct, clear, eminently didactic."

Francisco J. Ayala, Dept. of Ecology and Evolutionary Biology, University of California, Irvine (Cal.)

"As important as biology is today, most books in the philosophy of biology are too technical to be understood by students and the general public. In *Life Scientists*, Verschuuren gives his readers the benefits of his deep understanding of recent work in both biology and philosophy without burdening his exposition with unnecessary technicalities. Anyone who is interested in current thinking in the philosophy of biology will profit from reading this book."

David L. Hull, Dept. of Philosophy, Northwestern University, Evanston (Ill.)